PLACE, POLICY AND POLITICS

TITLES OF RELATED INTEREST

Accommodating inequality
S. Watson

City and society
R. J. Johnston

The city in cultural context
J. A. Agnew *et al.* (eds)

Cost-benefit analysis in urban and regional planning
J. A. Schofield

Gentrification of the city
N. Smith & P. Williams (eds)

Housing and labour markets
Chris Hamnett & John Allen (eds)

Housing and urban renewal
A. D. Thomas

Intelligent planning
R. Wyatt

Local partnership and the unemployment crisis in Britain
C. Moore, J. Richardson, J. Moon

Localities
Philip Cooke (ed.)

London 2001
P. Hall

London's Green Belt
R. J. C. Munton

Maps of meaning
P. Jackson

A nation of home owners
P. Saunders

New models in geography
N. Thrift & R. Peet (eds)

Place and politics
J. A. Agnew

The political economy of owner-occupation
Ray Forrest, Alan Murie & Peter Williams

The politics of the urban crisis
A. Sills, G. Taylor & P. Golding

Policies and plans for rural people
P. J. Cloke (ed.)

The power of geography
M. Dear & J. Wolch (eds)

The power of place
J. A. Agnew & J. S. Duncan (eds)

Production, work, territory
A. Scott & M. Storper (eds)

Property before people
A. Power

Race and racism
P. Jackson (ed.)

Remaking planning
T. Brindley *et al.*

Rural land-use planning in developed nations
P. J. Cloke (ed.)

Social theory and the urban question
P. Saunders

Technological change, industrial restructuring and regional development
A. Amin & J. Goddard (eds)

Town and country planning in Britain
J. B. Cullingworth

Urban and regional planning
P. Hall

Urban problems in Western Europe
P. Cheshire & D. Hay

Western sunrise
P. Hall *et al.*

PLACE, POLICY AND POLITICS

DO LOCALITIES MATTER?

Edited by

Michael Harloe

C. G. Pickvance

John Urry

London
UNWIN HYMAN
Boston Sydney Wellington

Published by the Academic Division of
Unwin Hyman Ltd
15/17 Broadwick Street, London W1V 1FP, UK

Unwin Hyman Inc.,
8 Winchester Place, Winchester, Mass. 01890, USA

Allen & Unwin (Australia) Ltd,
8 Napier Street, North Sydney, NSW 2060, Australia

Allen & Unwin (New Zealand) Ltd in association with the
Port Nicholson Press Ltd,
Compusales Building, 75 Ghuznee Street, Wellington 1, New Zealand

First published in 1990

British Library Cataloguing in Publication Data

Place, policy and politics: do localities matter?
1. Great Britain. Politics. Geographical aspects
I. Harloe, Michael II. Pickvance, Chris, *1944–* III. Urry, John
320.120941
ISBN 0-04-445505-4
ISBN 0-04-445506-2 Pbk

Library of Congress Cataloging in Publication Data

Applied for

Typeset in 10/11 point Bembo by Fotographics (Bedford) Ltd
and printed in Great Britain by Billing and Sons Ltd, London and Worcester

Contents

List of figures

Preface

The economics and politics of place have come on to both the political and social scientific agendas in Britain in the 1980s. As a result several programmes of 'locality' research have been undertaken, most financed by the Economic and Social Research Council (ESRC). The research presented in this book stems from one such ESRC interdisciplinary research initiative, *The Changing Urban and Regional System of the UK*. The project was carried out during 1985–7 and was based on work in seven localities (those reported in this book) as well as at the national level. We are immensely grateful to all our colleagues who worked on this initiative, to Philip Cooke – the project coordinator – to the ESRC and its officers and to the ESRC Steering Group chaired by Brian Robson. Thanks are also due to our editor at Unwin Hyman, Roger Jones, for his help and his patience with us! This is the second major publication to result from the initiative, the first being P. Cooke (ed.), *Localities* (Unwin Hyman, 1989). Others will follow.

In addition to the above acknowledgements we would particularly like to thank Paul Bagguley, Jane Mark-Lawson, Dan Shapiro, Sylvia Walby and Alan Warde of the Lancaster Regionalism Group; Keith Bassett, Martin Boddy and John Lovering of the University of Bristol; and Nick Buck, Ian Gordon and Peter Taylor-Gooby of the University of Kent. All of these contributed to the three chapters on Lancaster, Swindon and Thanet respectively and more generally to our work as editors in shaping and structuring this book.

Michael Harloe
Colchester
C. G. Pickvance
Canterbury
John Urry
Lancaster
July 1989

1

Introduction: the institutional context of local economic development: central controls, spatial policies and local economic policies

C. G. PICKVANCE

At a time when supranational institutions and international economic trends are having an ever-greater influence on our lives, and when local government is often thought of as more centralized than ever, it may seem pointless to write a book on the experience of seven localities. Our justification is not that we believe localities can be studied in isolation, but that broader processes do not fully explain what happens in particular places. Every locality bears the mark of the past and this produces distinctive effects on current processes. In addition, the current processes themselves interact in novel ways in particular localities. More generally, there is both academic and impressionistic evidence of local diversity: voting patterns continue to show indelible local effects, the competition for jobs between localities has led to the stressing (and creation) of local distinctiveness, and local councils pursue very different policies in many spheres. The chapters of this book are therefore committed to exploring local particularity as well as general processes.

Our aim is to present studies of seven localities in England in order to explore the range of policies they display towards the local economy and how these are related to their social, economic and political character. The seven localities were all studied between 1985 and 1987 as part of the Changing Urban and Regional System initiative of the Economic and Social Research Council.

In this introduction we examine the framework within which local policies are made. This is done in two stages. The first concerns the scope for local authorities to determine their own policies, and the second concerns the substance of central spatial policies, and of local authority exploitation of areas of policy discretion. In the first two sections, we confront the considerable body of writing that argues that local government has no scope for autonomous action because central control is so extensive. To the extent that this argument holds true, the scope for local policy variation disappears. Section 1 examines recent theorizing about local government and about its capacity for 'autonomous' policy-making, and section 2 considers recent government policies affecting the scope and character of local government. In section 3 we proceed to the second stage in our discussion by examining the development of central policies towards local areas and of local government economic initiatives. We leave until the conclusion of the book a discussion of the reasons for the differences in policy between local governments.

1 Theories of local government and its capacity for autonomous policy-making

In one sense the character of local government in Britain is obvious. District and county councils make up the lowest level of government and have particular responsibilities for education, housing, social services, planning, leisure, etc. They are financed by rates and government grants. Councillors are elected at local elections and councils are run by the largest party, which sets rates levels and decides how to spend money among the policy spheres for which the council is responsible.

Such a picture is, however, a formal description and neglects most of the questions that social scientists wish to ask about local government. How much effective freedom does local government have from central goverment control to follow the policies it chooses? How much is local government constrained by local economic conditions and by local pressure groups such as businesses or environmentalists? Do local elections allow voters an effective voice in council policies? In the spheres where councils have autonomy, what is the balance of power between politicians and professional 'officers' in making policy? How variable are councils in their delivery of services, and in their spending levels? Do differences in voting patterns reflect the past and present socio-economic structure of the locality?

Clearly the answers to these questions are interrelated. If local government is strongly controlled centrally then the possible impact of political parties, council officers, local pressure groups and local economic conditions is limited. Conversely, evidence of the strength of these forces is evidence of the weakness of central control. This is a point worth making when the image that local government is totally centralized is widespread. If central control were total there would have been no need for central government to have engaged in a continual battle to control local government spending, or to discredit 'loony Left' councils.

The main theories of local government are of very different sorts. There is a strong tradition of normative theorizing, which seeks to measure the extent to which local government matches up to ideals such as 'accountability', 'democracy', 'responsiveness', 'autonomy', or 'efficiency', and then puts forward recommendations to improve it. Examples of this can be seen in the work of the Maud Commission on Local Government in England (Cmnd 4040, 1969) and the Layfield Committee on Local Government Finance (Cmnd 6453, 1976). Intellectually, its roots are in public administration and welfare economics. Such theories play down the economic and political constraints that have shaped the system and that affect attempts to change it, and are not discussed below. The substantive theories, which will be discussed in this section, ask how local government comes to have the character it has and look for economic, social and political answers.

All theories focus on particular levels of question. The broadest theories ask why we have a system of local government (rather than an alternative such as decentralized administration), and why it has the range of functions it has. More medium-level theories ask why council policies take particular forms. We consider these in turn.

(a) Theories of local government and its functions: 'local state', 'dual politics' and 'uneven development'

The best-known of the broad theories are Marxist in inspiration. These are the 'local state', 'dual politics', and 'uneven development' models. The concept of 'local state' was introduced by Cockburn (1977), who used it to adapt Marxian theories of the state to explain local government in Lambeth. She argued that capitalism requires the performance of certain functions for its survival: the sustenance of capital accumulation (by the provision of infrastructure, the reproduction of the labour force, etc.) and the preservation of social order (by repression and ideology), and that these functions are performed both at central and local government levels. 'In spite of its multiplicity, however, the state preserves a basic unity. All its parts work *fundamentally* as one' (p. 47). She argued that the state must not be seen to be working in the interests of the dominant classes, so a 'detached integrity' is necessary (p. 48). Cockburn saw local elections as non-threatening to the capitalist order, because all the parties contesting them support that order and elections provide a legitimate form for the expression of social conflict. Despite her general emphasis on the unity of central and local state, Cockburn points out that the integration of the state is imperfect and there is some 'play' which provides an opening for working-class pressure. It is also, presumably, the failure of local government to fall in with governmental priorities that explains the need to introduce corporate management, a focus of her study. Cockburn's view was thus of the state (central and local) fulfilling indispensable functions, but also with conflicts of all kinds (worker v. capitalist, worker v. local government, service consumer v. local government and women v. men in the household) impairing its ability to do so. There is clearly a tension in this model, because it assumes that functions essential to capitalism will be performed, but then suggests reasons why they may not be performed.

The advantage of the 'local state' concept is that it theorizes local government as part of the capitalist order generally and insists on the linkage between central and local state. Its disadvantages are that it fails to explain why local government is necessary, or why it has particular functions, and that it presents an exaggerated picture of the unity of the state.

A theory that is less open to these criticisms is the dual state or dual politics thesis (Cawson and Saunders, 1983). This also starts from Marxian reasoning but, unlike Cockburn, argues that elected local government is necessary to help generate legitimacy for government. Governmental functions are classified as 'production' and 'consumption', and it is argued that the former will be placed at central level and the latter at local level. The assumption here is that production-related functions (e.g. industrial policy, labour relations, coping with unemployment) are too important to be left to local government and are best dealt with at national level through discussion between management, unions and government. This 'corporatist' style of politics contrasts with the 'pluralist' or 'competitive' style of politics said to characterize local government. The argument is that functions such as education, housing, social services and leisure are less important and can be opened up to the wider range of pressure groups found at the local level.

It should be made clear that Saunders does not claim that empirically in every country the two sets of characteristics he associates with central and local government will be found (i.e. production or consumption related activity, corporatist or pluralist politics, private property or public service ideologies, etc.). Rather he uses an ideal type methodology, which claims no more than that there is an elective affinity between the features mentioned and central and local levels of government and a *tendency* for them to occur together.

The dual politics thesis has generated a considerable debate (summarized in Saunders, 1986). However, much of it has missed the point because it has failed to address the theoretical underpinning of the thesis.[1] The key issue, in my view, concerns the validity of the theoretical arguments explaining the two patterns identified. Specifically, why should one expect local government to perform certain functions and be characterized by a particular type of politics (pluralist) and ideology (public service)? The answer seems to be that Saunders follows the Marxist claim that the state will operate more in the interests of the dominant class when production-related functions are concerned, and will be more responsive to popular opinion in the case of consumption-related functions. This is in line with the reasoning of Friedland *et al.* (1977) that government structures need to separate 'accumulation' and 'legitimation' functions. The implicit argument is that policy in support of accumulation or production needs to be concealed because it may generate political opposition, whereas the provision of services such as education, housing and social services will not. As O'Connor (1973) puts it:

> A capitalist state that openly uses its coercive forces to help one class accumulate capital at the expense of other classes loses its legitimacy and hence undermines the basis of its loyalty and support . . . The state must involve itself in the accumulation process, but it must either mystify its policies by calling them something that they are not, or it must try to conceal them (p. 6).

This reasoning and the assertion of a 'contradiction' between the accumulation and legitimation functions of the state is, however, in my view based on a fallacy. In most capitalist societies state support for production (e.g. subsidies to industry, infrastructure provision) is *not* popularly understood as a class-biased policy but is both publicly promoted and generally perceived as a policy in the national interest. Hence the need to conceal it from a supposedly class-conscious public is non-existent. Only in societies where there is a highly developed class view of the state would the alleged 'contradiction' arise.

The reasons given by Saunders for the placing of production-related policy at central government level are thus not convincing. His arguments regarding the placing of consumption-related policies at local government level are no more so: Saunders assumes that such policies can be placed there because they do not need to be concealed and are important in legitimizing the social order, but as we have seen this is dubious, or at best a partial explanation. Moreover, Saunders's argument that local politics is 'competitive' ignores much evidence on how local politics actually works. Dunleavy (1979) has mounted a strong case that local election outcomes are largely determined by national party popularity rather than by local council policy,[2] that local government is largely insulated from public opinion (rather than responsive to it) and that professionals are a key interest group affecting policy. Saunders (1983) has only more recently recognized professionals, and has "allocated' them to the regional state. Thus, local government cannot be identified with competitive politics as Saunders argues. Furthermore, the post-1979 demise of corporatism in Britain removes an important empirical prop for his conception of central government politics, and the vast range of national-level pressure groups suggests that competitive politics may characterize that level as much (or as little) as the local level.

We would conclude that something other than the alleged accumulation and legitimation functions of policies is needed to explain their location. A historical account of the development of state administration and local self-government would reveal the importance of a number of other considerations: notions of public interest (e.g. in public health); private property rights and citizenship rights;[3] economic arguments about the efficiency and cost of service provision and the non-duplication of functions; political arguments about the need for local democracy and a level of government that is not too remote from voters. It is out of this jumble, rather than any neat functional classification, that the distribution of functions emerges. Where Saunders is right, however, is in noting that the distribution of functions does have the unintended effect of setting the agenda of issues that can be discussed at each level of government.

Another Marxist-inspired theory of the local state is Duncan and Goodwin's (1988) 'uneven development' theory. The theoretical origins of their approach lie in the derivationist view of the state. This view (a) sees the state as performing functions necessary to capital that capital cannot execute itself; (b) 'reduces' the political sphere to these economic imperatives; (c) insists that the capitalist character of state intervention inheres in its form (e.g. relations with service users) as much as in its substance (e.g. support for capitalist firms) (LEWRG, 1979). Duncan and Goodwin depart from this approach in two ways: by their emphasis on uneven development, and by their rejection of economic reductionism. First, they argue that capitalist development is spatially uneven and that the need for a local level of government is a product of this: it is the only way central government can hope to manage the effects of this unevenness. Secondly, they argue that this unevenness is not solely economic: there is local variation in 'social relations' (such as class relations, gender relations and political traditions) which is not simply a product of the economic base. As a result, local government can fall into the hands of groups who (in extreme cases) reject the 'rules of the game' set by central government.

As a theory of the existence and structure of local government this approach is hampered by its functionalist origins. If the form of local government is to be explained by the need to manage spatial unevenness in economic development, then there should be a correspondence between the heterogeneity of local government structure and the extent of unevenness. But the main feature of English and Welsh local government structure is that, except in the largest cities, it is a uniform two-tier system. Indeed, compared with the French system of communes, whose population can range from a million to under one hundred, the English and Welsh structure is strikingly uniform (Ashford, 1982). The explanation thus fails to account for the observed pattern. Where the theory has more plausibility (although Duncan and Goodwin do not make this claim) is in accounting for the character of Scottish and Northern Ireland local government, to which the economic as well as political elements of territorial management are surely crucial.

As a theory of the distribution of functions between levels of government, Duncan and Goodwin's theory therefore has little to offer beyond the insights of O'Connor, Friedland *et al.* and Saunders. Its greatest value is as a theory of why local government policy varies between areas and its rejection of economic reductionism in explaining this variation.

In conclusion, the Marxist-inspired theorizing of local government seems misdirected. The more tenable explanations of local government and its functions are far more prosaic.

(b) Theories of local government policy-making and service provision

So far we have discussed theories that address the question of why local government exists at all and why it has particular functions. We now consider theories that seek to answer the question of what influences policy-making by local government, its variability between areas, and its responsiveness to central government pressure. These are of two kinds, and are complementary to one another. The *structural* theory points to the particular character of central government/local government relations in Britain and the scope it allows for local discretion; and *local* theories explain the use made of this scope as reflected in the influence of local socio-economic conditions, local pressure groups, etc.

The structural theory starts from the fact that, comparatively speaking, Britain belongs to the category in which central government is 'non-executant' and local government 'executant' (Sharpe, 1979). Central government makes policy, but leaves implementation to local government within a framework of legislation, departmental advice and other controls. The majority of expertise needed to execute policy is possessed by professional staffs employed by local authorities. The contrast is with systems like the French where certain central government departments have a network of local offices, which employ professional staff who not only advise on but also carry out local policy. In such cases local government employs fewer staff – 11 per cent of all public employees compared with 38 per cent in Britain (Page and Goldsmith, 1987, p. 157).

This structural feature has two major consequences. First, it means that local government in Britain possesses important resources: as an elected level of government it has political legitimacy; it has responsibility for service implementation and hence possesses resources of expertise, staff, land, buildings, etc.; it has the right to levy rates or poll tax and service charges; and it has the information that comes from its 'coal-face' role. Each of these has been deployed in recent conflicts with central government, as will be seen below.

The second consequence is that the legislation under which local government works allows it considerable discretion. The fact that legislation authorizing local government to perform certain functions is often vague about the precise character of the services to be provided or about the level of provision gives it much room for manoeuvre. Legislation is usually classified as mandatory (that which obliges councils to provide certain services) and permissive (that which allows them to do so). An extreme example of mandatory legislation is the 1980 Housing Act, which gave council tenants the right to buy the houses they occupied at substantial discounts. At the other pole is a measure that is important in local authority economic development policy: Section 137 of the 1972 Local Government Act gives councils a blanket right to spend up to the product of a 2p rate on any purpose beneficial to part or all of the population of its area. (From 1990, with the abolition of domestic rates, this will be converted to a per capita sum.) Services subject to permissive legislation such as the provision of parks, libraries or industrial estates allow councils discretion. However, the permissive/mandatory distinction is not absolute, because even mandatory functions such as education leave great scope for councils to decide how to provide it, e.g. via comprehensive or grammar and secondary modern schools, with what class sizes, and what curriculum (Crispin, 1983). The vagueness of most legislation bearing on local government can be seen from the fact that legal cases brought by parents and others, alleging failure of councils to provide an adequate service owing to cuts in spending (e.g. that education included too narrow a range of subjects), have uniformly failed. The legislation turned out not to specify any required level of provision.

If the resources and legislative discretion possessed by local government under an 'executant' type of system explain why variation between councils in service provision is possible, the remaining theory seeks to explain what use is made of this room for manoeuvre. It focuses on the various local 'conditions' (social, economic, political) which appear to affect council provision – these are mostly derived from correlation and regression analysis of service outputs – and on the local actors or pressure groups whose influence usually emerges in case studies.

By definition, output studies cannot uncover the influence of factors affecting all British local authorities such as the structural factors just discussed. However, they can point to conditions differentiating them. Unfortunately most of the studies are rather dated. The best of them (using pre-1974 data) shows systematic variation in spending levels by type of authority (e.g. suburban authorities, county towns, seaside resorts), by position in urban hierarchy and by location in Wales (Sharpe and Newton, 1984). More generally, a major finding is that Labour party control is correlated with higher spending on services such as housing, education and social services, even after measures of need are controlled, whereas it does not affect services such as fire, refuse and amenities or capital spending on houses and roads (which were correlated with need measures) (Sharpe and Newton, 1984; Pinch, 1985).

Output studies have their limitations; for example, they rely on quantifiable variables, and the level of explained variation is low (Pinch, 1985). It is in case studies that the influence of individuals and groups can be seen more clearly. An explanation frequently put forward for high spending levels is that it is due to the influence of professionals, especially in authorities which, owing to long-term control by one party, have allowed professionals to have more influence (Cheshire, studied by Harrington, 1984). The argument is that whereas, constitutionally, local councillors take the decisions on service provision, they do so on the basis of choices offered by senior officers in each service-providing department. These officers can exert significant de facto control of policy within the area of statutory discretion, for three reasons: as permanent employees they are in a strong position to mobilize support for a 'departmental' view that may conflict with politicians' preferences; as professionals they can claim technical expertise; as members of nationally organized professional associations they can appeal to ideas of 'best practice' prevailing in the profession (Pahl, 1975, Part III; Dunleavy, 1979).

The role of professionals currently is a matter of debate. Changes in management style, due to the introduction of corporate management in the 1970s and in the face of cuts in spending, may have reduced the power of professionals generally. The increased politicization of local government and the increased number of hung councils (20 per cent in 1985) has reduced this role further (Leach *et al.* 1986). The possibility of professionals appealing to national norms has been further weakened according to Sharpe (1985), who argues that professional associations themselves have been co-opted by central government (see also Stoker, 1988). Another emphasis appears in the work of Laffin (1986) and Rhodes (1988), both of whom point to the differing power positions of different professions. Laffin shows that highway engineers are well-placed in both central and local government, that there is an interchange of staff between levels and a well-developed 'professional community'. By contrast, housing management is a semi-profession, which has no representation at central government level and is in an altogether weaker power position.

Evidence on the ability of pressure groups to influence local government is mixed. Against the image of an open pluralistic distribution of power in which groups could form easily and hope to influence council policy, a major theme of the literature has been to emphasize the insulation of British local government from external pressure and especially from pressure from residents (Elkin, 1974; Dunleavy, 1979). This is seen as the counterpart of an elitist, party-dominated and secretive style of government in which the interests of the public are best served when pressure groups are ignored. (In the USA the reverse extreme applies – the public interest is seen as best served when demands from pressure groups are met.) This 'elitist' model allows for considerable business influence either involving direct pressure, or through shared values, as when the council assumes it knows what business priorities are. It also acknowledges that residents' influence increases somewhat prior to key elections, or when residents are demanding the cancellation of a project that could benefit other spending departments. A recent suggestion, however (Stoker, 1988), is that this model was valid in the 1970s but is now out of date owing to the 'opening out of local authorities' (due to greater councillor openness to local interests, the introduction of neighbourhood levels of management and a greater community orientation among council professional staffs) and increased local group activity (including willingness to help run services). Stoker also argues that different types of authority are receptive to different types of pressure, e.g. left-wing Labour authorities to minority and cause groups, New Right authorities to business and middle-class residents' groups, etc. These are all hypotheses that remain to be tested.

The implications of this debate for local council economic development measures depend on the extent to which they have negative impacts on local residents. If they do not, policy is likely to be made by local councillors and council officers, with more or less consultation with local business interests. If local residents are affected then policy may or may not be modified.

The argument so far is that local government in Britain is executant in character and this means it has a considerable number of resources and hence discretion. It is because of this scope for discretion that party control can make a significant difference and that groups such as professionals, business people, ratepayers and environmentalists can have some influence on policy. Their success, however, depends upon the local authority's capacity to insulate itself from such local pressure.

Theories of local government policy-making and service delivery thus refer to the structural character of local government (executant, with all the resources that entails), to the scope for party control, to professionals and to external groups. All these will be referred to in the chapters in this book. Local economic policy may be a response to external pressure, but more often it is the council's role as legitimate representative of territorial interests that gives it the initiating role.

2 Recent central government policy towards local government

So far we have approached the question of the autonomy of local government by exploring its structural character and degree of openness to local processes. We now continue our answer by examining recent government policy towards local government. This will allow us to ask whether recent 'centralizing' policies have reduced or eliminated the scope for discretion discussed earlier, and indeed what the future shape of local government is likely to be.

Historically, the tradition of non-executant central government and executant local government is linked with what Bulpitt (1983) calls the 'dual polity'. This refers to the arrangement, which prevailed between 1926 and 1960, by which policy is divided into 'high politics' (issues such as defence, national economic policy, foreign policy) and 'low politics' (issues such as local government and service delivery). Bulpitt argues that central government was content to deal with 'low politics' by leaving local government to get on with service delivery as long as it kept services out of the national political arena. Bulpitt's thesis is that this arrangement worked because of the consensus that had been established between central and local government. He rejects the idea that it involved the imposition of central control, on the grounds that the localities showed no will to resist: local elites were fundamentally cooperative. The period since 1960 has seen a breakdown of the dual polity. Partisan politics has completed its penetration of local government and the impact of recession on the welfare state has meant that local government spending has come under attack as a crucial element of public spending (Ashford, 1980).

The crisis phase in central–local government relations dates from 1976 when the Minister responsible for local government declared 'the party's over' and that local government spending needed to be controlled. Since then there have been continuous attempts by the centre to control local government spending. Whereas Labour's 1976 cash limits policy was intended to control local government spending as an element within public spending, Conservative governments have had additional aims: to weaken producer (e.g. trade union) influence in local authorities, to increase accountability to the public and service consumers, and to prevent the use of local government by the 'new urban Left' as platforms for opposition to national government. This has led to policies that include encouraging the contracting out of functions to private enterprise and setting up rival institutions to carry out local government functions, and to attempts to increase the influence of parents, tenants and voters at the expense of teachers or council officers. Paradoxically, the pursuit of this increased freedom has often involved new central controls. Clearly, central–local government relations are in transition from the certainties of the 'dual polity' to a situation where the centre seeks to remodel an executant system of local government that retains a considerable capacity to resist.

The measures taken in this direction now will be outlined, together with their effects.[4] Have they been successful or have the resources of local government indicated earlier allowed successful resistance?

Controls on spending. The first measures introduced in 1981 sought to encourage councils to curtail their *current spending* by setting targets, supported by penalties if the targets were not met. The penalties took the form of reduced central grant. The targets were based on central assessments of local authorities' 'needs'; councils, however, protested that targets were simply a way of dividing up a notional aggregate target that bore no relation to need. The targets were manipulated partly to hit Labour councils and partly to avoid too drastic effects on Conservative councils. But their effect was very limited and local government current spending increased at 0.7 per cent per year in real terms between 1979 and 1987 (CSO, 1988[5]). The failure[6] was due to four factors: (i) the unwillingness to apply severe targets because of the opposition it would have engendered from councils of all colours; (ii) the continued increase in real terms of central government grants to councils – technically, by cutting grants, a severe squeeze could have been imposed on councils, but this was not done to avoid opposition;[7] (iii) councils' ability to increase rates levels to make good any reduction in central grants (this was partly due to a determination to maintain spending and partly a precautionary measure to cope with the uncertainties of income due to the system of penalties); (iv) creative accounting by councils, which allowed them to minimize the effect of the controls. This included techniques such as shifting revenue items to capital account, manipulating reserves, postponing debt repayments, mortgaging assets and deferring purchases (Parkinson, 1986).

Controls on income. The failure of controls on current spending (which accounts for some 90 per cent of all local government spending) led government to consider additional control methods. One of these was the setting of limits to the rates that councils could set, known as 'rate-capping' (under the 1984 Rates Act). The raising of rate levels was an important escape route by which councils could avoid government attempts at a financial squeeze (though as we have seen these attempts were half-hearted). For example, the Greater London Council had financed its policy of low fares on public transport by a large increase in rates to make good the loss of fare income. The consequence was that it lost all of its government grant that year (the maximum penalty). Once this sanction had been applied, and ignored, the government had no further means of influence. Removal of the right to set rates was therefore an obvious solution.

The right to set rates was a jealously guarded prerogative, and the threat to it led to widespread council opposition during the passing of the legislation. Like the system of penalties, rate-capping is technically a measure that could have enabled strong central control on council revenues. But in practice the number of rate-capped councils has only been between ten and twenty, and even among these rate increases have often been allowed. The threat of rate-capping may have had a deterrent effect in a larger number of local authorities. The effect of rate capping has thus been relatively small in aggregate, but significant in some of the rate-capped councils. The effects in 1985–7 were mitigated by central government's ignorance of the true financial situation of councils, but this is now less true. Again, the government has to steer a way between an extensive use of this power (which would mean dealing with many recalcitrant councils) and a limited use to avoid trouble. Its effects have been much less than expected.

Abolition. A further government response to the 'high spending' of certain Labour councils and to their use of local government as a platform for opposition was the abolition of the Greater London Council and six metropolitan county councils in April 1986. Though cloaked in the language of avoiding duplication of functions and saving money, no one (including leading Conservatives) doubted that the real motive was political (Pickvance, 1986a; King, 1989). This measure was the major achievement of the government in its policy toward local government.

In view of the failure of most of its other attempts, this success needs explanation. At a political level, opposition was less than in the case of rate-capping. Some Conservatives were sensitive to the democratic argument that abolition would appear as a substitute for a failure to win power through the ballot box, but there was widespread support for the measure. Moreover, Labour was divided, because many district councils had opposed the creation of metropolitan authorities and stood to regain some of their former functions (e.g. cities such as Liverpool, Manchester and Birmingham).

However, the abolition measure can also be seen as the latest in a long line of drastic changes to the territorial structure of local government in Britain, which Ashford (1982) attributes to the highly centralized British system of government. He makes a contrast between the French system, where local interests have easy access to central government and can thus prevent changes that they do not support, and the British system, where local government has been kept politically (though not administratively) isolated. The weaker representation of local government interests at the centre in Britain, which was part of the dual polity model, is the condition facilitating the drastic reorganizations we have seen. This suggests that the end of the dual polity has not led to any gain in political strength for the localities.

Weakening local government via contracting out and privatization. A further attack on local government has been through the 1980 Housing Act, which gave council tenants the right to buy their houses at a discount (privatization) and the encouragement of tendering and thence contracting out of council functions. The sale of council houses, billed as 'returning power to the citizen' and 'helping dismantle council bureaucracy', has been a major government success. It met little council opposition (Forrest and Murie, 1985), and Labour policy has largely fallen into line. Contracting out has so far been on a small scale. It has been voluntary and has usually occurred in 'housekeeping' services such as refuse collection and school cleaning. Councils have been reluctant to contract out services and Ascher (1987) describes the debate about the issue as 'a storm in a teacup' (p. 227). Many councils fear loss of day-to-day control of a service, and the deterioration of quality that is often associated with contracting out, as well as worsened industrial relations. However, current legislation *obliges* councils to put more services out to tender. It is notable that the services concerned are those employing manual workers, i.e. where trade unions are strong. Services employing mainly professionals have not yet been threatened by compulsory tendering. In general, reform of the professions is proving one of the most difficult tasks for the government.

Weakening local government by diminishing its functions and setting up rival institutions. The failure of three out of the four types of initiative discussed above has led to a new and more ominous series of initiatives. As we shall see in Section 3, since 1979 Enterprise Zones, Urban Development Corporations and various inner city initiatives have been launched, which have in common an intention to by-pass local government. However, arguably these do not involve key council functions or where they do have been subject to compromise. For example, councils have retained planning control over medium to large shopping developments in Enterprise Zones.

The latest initiatives, however, are more radical in intent and tackle major functions. In education, the largest council spending item – local authority control of schools – is threatened by proposals to decentralize management and give parents the right to vote schools out of council control. (Their funds would then come directly from central government.) City Technical Colleges, reliant on business funding, are also a threat to local education authorities, and control of polytechnics passed from local government under the 1988 Education Act. In council housing, no less radical changes are proposed for those tenants who do not wish to buy. A choice of landlord is to be offered to tenants, and Housing Action Trusts are to be created to supplement the choice between housing associations and councils.

Although few of these initiatives have yet made much progress, underlying them is a model of local government as one supplier of services in competition with others (rather than a monopolistic supplier in many spheres) and as a budgetary centre receiving rates (or poll tax) and central grants and making contracts with private and other contractors to provide services, but employing few executive staff itself.

Poll tax. The final measure to be considered is the abolition of domestic rates in favour of poll tax. This took place in Scotland in 1989 and will be introduced in England and Wales in 1990. Its aim is to make councils more responsive to voters. In place of property-based domestic rates, the poll tax will be a per capita charge on adults whose names are on a specially drawn-up register. It will be calculated to meet the gap between budgeted spending and income from central grant and centrally distributed non-domestic rates, and will be subject to a 'cap'. Non-domestic rates will remain property-based, but will be set nationally at a uniform rate and then distributed to localities. Ironically this will remove business concern with local taxation. The redistributive effects of these changes are considerable: small households and those living in expensive houses benefit at the expense of large households and those living in cheaper houses. The increased 'accountability' is intended to come through the fact that, whereas at present councils levy rates on voters and non-voters (i.e. payers of non-domestic rates), in the future councils will only have the power to tax voters. This will increase the impact of spending decisions on voters and hence make them more 'sensitive' to the council's policies. This is because central grant and non-domestic rates income will be centrally fixed. The burden of extra spending must thus be borne entirely by voters instead of being shared between domestic and non-domestic ratepayers as at present. This is intended to shield business from 'high-spending' councils and pass on the whole effect of increased spending to households.

Between 1979 and 1987 councils' rates income increased at 3.1 per cent per year (CSO, 1988) and proved the main means by which councils maintained spending levels in the face of small real increases in government grant. If poll tax blocks this source of relief from central financial squeeze, it will be the first ever successful central attempt to control local finances. It is more likely that it will fail, because voters will continue to tolerate 'high' spending, or 'support' it indirectly by treating local elections as polls on national party popularity or because swings in voting do not lead to changes of council control. It is also likely that central grants and non-domestic rates levels (and hence yields) will be fixed at 'generous' levels from the point of view of councils, thus easing the burden on poll tax payers. In this way the risk of massive voter dissatisfaction (which could be expressed at a general election as well as at local elections) could be averted. It is thus too soon to say whether this measure will change the ground rules of central–local government relations.

What then is the state of the battle between central and local government?

We can start from the paradox that central government has found it easy to change the territorial structure of local government through the abolition of the GLC and MCCs, but when it is a matter of control of spending or income-raising it has so far failed. This is all the more striking because its chosen methods, the withholding of central grant to discourage spending and the rate-capping power, are *technically* powerful enough to achieve almost total control of local government finance. What, then, has prevented them from being used to greater effect?

The answer lies in our earlier discussion of the dual polity and the executant character of British local government. We would argue that, although the dual polity has led to a denial of central political influence to local government, the executant character of British local government (particularly owing to its major welfare state functions and economic impacts) has given it an indelible administrative power. The evidence of local government's use of bargaining resources in this section supports our earlier claim that local government has some key resources due to its interdependence with central government: (i) its concentration of expertise and staff and 'coal-face' role in local service delivery has made it fairly difficult for central government to contemplate direct control of rebel local authorities. As Ashford (1982) writes, 'members [of the British higher civil service] probably could not perform most of the tasks of local government' (p. 9). Its knowledge of the local financial situation has facilitated creative accounting and given councils a weapon against central government: (ii) its legitimacy as an elected body and hence public support for the right to set spending levels and rates levels in defiance of central wishes and avoid the squeeze on central grants, and (iii) its reliance at the polls on the local party machines, which the governing party relies on at general elections. The reliance of the Conservative party nationally (e.g. to re-elect MPs) on support from local party organizations has meant that local Conservative opposition has been heeded, thus tempering central government policy.

This analysis explains why central attempts to control local government up to 1987 have been so unsuccessful, and why abolition has been the major success. Local government resources are most important in its day-to-day working and make control of this difficult for central government. By contrast, they offer little protection against structural change to the system, e.g. territorial reorganization. The paradox noted earlier is thus resolved: it is quite consistent to argue that central government is both powerful in making some kinds of change (structural reforms) but weak in others (policy implementation).

The relative success of councils in resisting central policy up to about 1987 seems unlikely to continue. Some of the previous solutions are no longer viable: creative accounting allowed some one-off changes in accounting practice and pushed financial problems into the future – but they are now having to be faced. Raising rates to finance spending will be transformed into raising poll tax levels – which for the reasons mentioned will be more difficult but not impossible.

However, the major reason for believing that central–local government relations are entering a new era is that for the first time the centre is intervening in the provision of services, rather than in their financing. The introduction of compulsory tendering[8] is an attempt, in the name of cutting costs, to facilitate private enterprise provision of local government services (whether mandatory or permissive). A second initiative consists of the renewed attention to efficient management led by the Audit Commission with its regular reports on comparative costs of service provision among local authorities. The key point about both these forms of intervention is that they bear on the coalition of local politicians, local government professionals and manual workers (and their unions), which has been the backbone of local government resistance. The Conservative government has correctly identified the obstacle to its previous efforts to control local government and is seeking to remove it.

We are thus in a transitional situation. The government has failed to control the operation of the executant local government system that has existed until now. It has thus been obliged to change that system. What is uncertain is how far the government will have to intervene to change it, and whether the changes introduced will themselves have unintended effects which negate the intentions behind them. The major risk following from the transfer of services to private contractors, the transfer of housing to housing associations and Housing Action Trusts, and the devolution of management to schools is that these functions become more difficult to manage or give rise to new problems without mechanisms for solving them because of the indirect relationships involved, and also become less likely to keep to their budgets. One scenario would portray efficient management (and evenness of service provision) and adherence to spending targets as obsolete aims, and as prices worth paying to end the executive role of local government. Another would envisage public reaction to inefficient management and costly provision as being sufficient in scale to lead to a partial reversal of the changes. Everything will depend on the balance of power between central government, producer organizations (local government trade unions and professional associations), voters and local politicians.

If local government is to become more than a budgetary centre receiving grants and poll tax and issuing contracts for services, it will need to develop new capacities in fields that do not involve service delivery and where it has special legitimacy as a territorial authority. Planning and the shaping of the locality's economic and social character are two such functions. One of the subjects of the next section, local economic intervention, is thus a function that is likely to remain important even on the most extreme visions of the future of local government.

3 Local and regional economic policy

The economic fortunes of particular localities have always depended on a combination of local, national and international economic activities. However, this century has witnessed major changes in beliefs about the desirability of state intervention in the economy, about the value of national policies favouring particular places (whether on the scale of regions or inner cities) and about the capacity of local councils to take an active part in developing their own economies. Hence the role of the public sector in local economies has expanded enormously.

In this section we shall outline the final part of the institutional context affecting the economic fortunes of localities. We shall describe central government policies for local areas (and in particular regional policy) and the growth of local council economic development powers.[9] To start with, we discuss the concept of local economic intervention and some of the precursors of present policy.

(a) Origins of local economic intervention and regional policy

The concept of local economic intervention is far from clear. In the broadest sense every economic activity of a local council that has a local effect – from the employment of a local resident to the purchase of supplies from a local firm – forms part of a council's local economic intervention. Likewise the outputs of a council's services such as education or housing contribute to the creation of a local labour force and hence to the local economy. If we restrict the concept to interventions that are clearly intended by actors such as councillors to promote the local area, the situation is hardly clearer. Many features of a council's services, such as the quality of its schools, parks, leisure facilities, public buildings, etc., may be intended to gain a competitive edge over neighbouring areas and attract residents, tourists and employers. Even if the term is restricted to interventions that intentionally benefit particular industries there are still many borderline cases. Theatres and leisure facilities may be crucial to a resort's ability to attract tourists and hence can be seen as support for the tourist industry, but they are equally facilities that are used by local residents.

In practice, most descriptions of local economic intervention define it in a very specific way: to refer to activities aimed at improving the manufacturing base of a locality (particularly by attracting new plants). Only recently has this definition been widened to include attempts to attract services sector activity and the efforts of councils to develop their own purchasing power and role as employers.

Local action to promote the local economy in the broadest sense has a long history in Britain. Its origins lie in the nineteenth century notions of civic improvement, municipal enterprise and municipal socialism, and in Keynesian ideas of state economic intervention.

The mid-Victorian period was characterized by urban improvements of all kinds. The provision of basic infrastructure, such as water supplies, sewerage and railway services, was a result of a mixture of public and private initiative. The building of town halls, public libraries, museums and colleges, and the planning of parks was a product of both municipal initiative and philanthropic endowment. These developments were costly and the speed at which they were implemented varied greatly between cities, depending on municipal economic fortunes, pro-spending political ideologies (such as what Briggs (1968) calls the 'civic gospel' in Birmingham under Joseph Chamberlain), and opposition from anti-spending interests. Likewise, business fortunes and philanthropic inclinations were varied. But however uneven the development of urban improvement, it demonstrated that councils had the capacity to increase the amenities of their towns and cities.

The emergence of a municipal role in civic improvement was not uncontroversial, however, and by 1900 political conflict around the issue was considerable. Various terms were used. 'Municipal enterprise' was the least controversial: it referred to the fact that in certain spheres municipal provision was widely believed to be more efficient. Less agreement, however, surrounded the extension of municipal activity under the banners of 'municipal trading' and 'municipal socialism'. Municipal trading was less controversial than municipal socialism because it represented a gradual outgrowth of municipal enterprise. Municipal socialism, however, was a Fabian notion born in the 1880s and involved radical changes in the scope of local government (Kellett, 1978). It envisaged councils acting as model employers themselves to offer jobs to the unemployed and setting wage rates and work conditions for their employees as an example to private employers (Buck, 1981). Councils would also extend municipal enterprise into spheres that were previously the preserve of the private sector, e.g. subsidized council housing. There was a pragmatic element in Labour Party support for municipal socialism, since between its formation in 1900 and entry into government in 1924 its status as a new party meant it had no choice but to seek success at local government level as a step towards winning national power.

Municipal socialism was important in the development of local economic policy because it involved a more explicit concept of a local economy (particularly a local labour market) on which the council was intervening. By contrast, in civic improvement and municipal enterprise this concept was at best implicit. But the prevailing belief before 1914 was that the economic experience of localities depended on the national economy. The nineteenth century saw the gradual growth of central grants in support of municipal activity, but this was to support service provision rather than ensure local economic prosperity. The concept of central government intervention in support of local economies was non-existent.

A second origin of local (and especially regional) economic policy is the spread of Keynesian ideas about the possibility of state regulation of the economy. The limited measures adopted in the 1930s in the depressed regions were a practical rather than a theoretically based response to social distress. But they acknowledged a government responsibility for the economic fortunes of localities, which was a novel step. The setting up of the Barlow Commission on the Distribution of the Industrial Population in 1937 was further evidence of the growth of this government responsibility. Many commentators see the 1940 Barlow report (Cmnd 6153) as a high point in the systematic analysis of spatial economic development, because it makes an explicit link between high unemployment in depressed areas and congestion in the South-East. Its proposals to assist industrial location in depressed areas necessarily involved measures to control expansion in the South-East. There is some debate about whether post-1945 regional policy, which followed closely the Barlow report's proposals, owed more to the 1930s Special Areas and the Barlow report or to the wartime experience of government economic intervention (Parsons, 1986). In any event, by 1945 both major parties were committed to the need for economic regulation to secure full employment, and this provided the political basis for the continuity of regional policy, despite changes in government, until 1979. The stagflation of the 1970s shook Labour faith in Keynesian methods of economic management, but this had no repercussions for regional policy until 1979, when the Conservative government explicitly rejected Keynesianism and the inter-party economic consensus of the post-war period.

By 1945 the conceptual basis for local and regional economic intervention was thus in place. In the rest of this section we discuss its development.

(b) The development of regional policy and local economic policy

(i) 1930–9

The most recognizable antecedents of regional policy and local economic policy as we know them today date from the 1930s. The Special Areas Acts between 1934 and 1937 allowed loans and tax incentives for businesses in the Special Areas, established trading estates through the creation of Industrial Estates Corporations, and continued a 1928 scheme to encourage labour migration from the depressed areas. This represented a substantial range of policy measures, of which the effect was only starting to be felt in 1938 when the rearmament drive ended high regional unemployment (McCrone, 1969).

The period also saw the first local council economic development initiatives. Councils gained the power to advertise themselves in 1931 (under the Local Authority (Publicity) Act) and before 1939 a large percentage of county boroughs, municipal boroughs and urban districts in England and Wales belonged to development organizations concerned with tourism and industry. Liverpool and Jarrow appear to have been the first councils to secure powers to build factories, develop industrial estates, and make loans to industrialists – under private acts in 1936 and 1939 (Camina, 1974).

The post-war period can be divided into three phases from the point of view of national regional policy and local economic policy: 1945–59, 1960–74, 1975 to date.

(ii) 1945–59

This period saw the initiation of a regional policy to steer manufacturing investment into development areas. The main method used was a combination of stick and carrot. Following the analysis of the Barlow report, firms would be offered the incentive of grants or loans if they located or expanded in assisted areas; expansion elsewhere was discouraged by the need to obtain an Industrial Development Certificate (IDC). In addition, funds for advance factories and land reclamation were available. The policy was pursued 'actively' up to 1950 in the sense that the level of grants or loans per project was high and IDCs were difficult to get outside assisted areas. After that, regional inequalities declined as the post-war recovery penetrated even to the regions that had been depressed in the 1930s, and regional policy went into a 'passive' phase. (For more detailed information on regional policy in this and later periods, convenient sources are Keeble (1976), Martin (1985), Martin and Hodge (1983), McCrone (1969), Pickvance (1981 and 1986b) and Townsend (1980).)

This period was one of New Town expansion, when the first generation of New Towns were created and grew. But of the thirteen New Towns designated between 1947 and 1950, no less than eight were overspill towns for London (Aldridge, 1979). The effect of building them up was that firms could locate in the South East rather than in the assisted areas. Thus, New Towns policy was in partial contradiction to regional policy. The 1950s saw the expansion of the first wave of New Towns, but the creation of only one additional one. Conservative policy favoured expanded towns – of which Swindon was one.

The 1945–59 period was also one where the pre-war foundations of local council economic policy were built upon at a steady rate. By the late 1950s, therefore, most of the present-day institutional machinery affecting local economic development was in place: national regional policy, local and regional development organizations and local council economic policy.

(iii) 1960–74

Regional policy. Only in 1958 did the first signs of a regional problem start to re-emerge. As the symptoms developed so regional policy was reactivated, and 1960–74 is generally seen as the period of highest regional policy activity. Incentives increased, IDC control became tougher, and the coverage of assisted areas expanded. The main methods adopted remained the offering of incentives to manufacturing firms to set up in development areas, and central government control over the opening of factories in the rest of the country. But there were four innovations. First, Regional Employment Premium, a labour subsidy to cut the wage bill of all manufacturing plans in assisted areas, was introduced in 1967 and remained in operation until 1976. Secondly, a new form of investment incentive was introduced in 1972. Whereas previous incentives had taken the form of grants accorded automatically to eligible projects, Regional Selective Assistance was discretionary and allowed a topping up of the automatic Regional Development Grant to offer a high level of incentive. Thirdly, a differentiated set of incentives was introduced by the creation, in addition to Development Areas, of Special Development Areas (1967) and Intermediate Areas (1970). The Intermediate Areas followed the recommendations of the Hunt Committee, which resulted from lobbying by Yorkshire and Lancashire. The introduction of SDAs was the first acknowledgement of differential economic experience *within* regions – a theme that later became very significant.

Finally, the coverage of assisted areas increased considerably. Between 1945 and 1958 the assisted areas consisted, broadly speaking, of the Special Areas of the 1930s (parts of South Wales, the North-East, Cumberland and Clydeside) plus the Scottish Highlands, and scattered areas in North-West England. From 1958 to 1966 numerous small areas were added to this set of compact regions: parts of North-West Wales, the South-West, Kent (including Thanet), the Yorkshire and Lincolnshire coasts and Western Scotland (Keeble, 1976). This was due to a change in the basis of designation. The Development Districts of 1960–66 were designated when the unemployment rate in an area reached a 'trigger' level, e.g. 4.5 per cent. This allowed a 'fine-tuning' of assistance, reflecting a 'social welfare' notion of regional policy rather than a concern with an area's potential for economic development. But it led to unpredictability and was abandoned as economic rationales overtook the social rationale and the regional 'growth pole' notion gained credence (see below).

From 1966 onwards, however, the 'patchy' assisted-area map changed. By 1970, with the addition of Intermediate Areas, the whole country North and West of the Midlands together with Devon and Cornwall had been designated as assisted areas, and this continued until 1979. (The gradations within the assisted areas were as follows: the Special Development Areas coincided largely with the pre-1958 assisted areas, and the Intermediate Areas were mostly in Yorkshire and Lancashire.) The justification for this wider coverage was that regional problems were not problems of particular localities but were region-wide. Hence solutions should be region-wide.

By the 1960s the rationale for regional policy was very different from the 1930s. Instead of being a social welfare measure, regional policy had become linked to growth policy and demand management: by spreading economic activity into all regions, it would reduce labour shortages, inflationary pressures would be kept down and faster growth achieved. This shows clearly the impact of Keynesian ideas. Region-wide solutions were adopted following the rise of the 'growth pole' model. According to growth pole theory an industry generates a network of links with suppliers and customers in the same region, and hence is capable of regenerating a regional economy.

As well as the revival of regional policy the 1960–74 period saw a new phase of New Town creation. Between 1961 and 1970 thirteen New Towns were created. Compared with the first wave of New Towns, the second wave was more compatible with regional policy as eight were located in assisted areas. The others were near Birmingham (two) and near London (three). There was an initial phase of dispersal of government offices, and the entry of the United Kingdom into the European Economic Community in 1972 opened up access to a new source of regional aid which was to prove significant in the late 1970s and 1980s.

Local economic policy. The 1960–74 period also witnessed a growing concern with economic development by local councils. A survey in 1971–2 revealed that most councils provided land for industry, somewhat fewer were engaged in promotional advertising, about one-third had industrial premises available and rather fewer offered financial assistance (Camina, 1974). A majority also belonged to local or regional development bodies. The significance of regional inequality is shown by the fact that, in each of these types of intervention, councils in development areas were more active than those outside them. The main legislation involved at this time was permissive, as it remains today. Councils could provide industrial land and buildings and offer subsidized mortgages for industrial premises built on land owned by the local authority (under the Local Authorities (Land) Act 1963). They could develop land for industry on their own or jointly with private developers (under the Town and Country Planning Act 1971) (Mills and Young, 1986). From 1963 local authorities were also given a general power to spend the product of a 1d rate[10] on anything of benefit to the area and its residents that was not otherwise authorized (Local Government Financial Provisions Act). Many used this in support of industrial promotion.

In 1972 a 'great leap forward' started in the extent of local economic intervention. Local authorities were allowed to form development companies and offer subsidies on the rents of industrial premises they owned (under the 1972 Local Government Act), and under section 137 of the same act the 'free 1d rate' became a free 2p rate.

The 1966–74 period was also notable for the initiation of a number of central interventions in inner city or other deprived areas. Unlike later interventions, local authority involvement and support was usually a condition of the schemes' going ahead. They were generally low-cost, short-term, 'experimental' projects and owed a lot to US experience. They include: Educational Priority Areas (1967), Urban Aid (1968), the Community Development Projects (1969), the Urban Guidelines Studies (1972), Comprehensive Community Programmes (1974), the Inner Areas Studies (1974) (Lawless, 1986). They shared a definition of the inner city problem as social and environmental rather than employment-related; the racial element was usually present but undeclared. Solutions were looked for in improved council service provision. Their small area focus, however, is due less to the geographically restricted nature of 'inner city problems' than to administrative convenience, the symbolic value of numerous small schemes and the radical implications of admitting the much wider incidence of many 'inner city problems' (Berthoud, 1976; Eyles, 1979; Hamnett, 1979).

(iv) 1975 TO DATE

Regional policy. The period since 1975 has seen the onset of recession and unprecedented levels of unemployment. Regional policy has continued, but underwent major changes in 1979 and 1984.

By the 1970s regional policy had become a matter of incentives only: the element of constraint on firms locating or expanding in affluent regions (via IDC control) had been effectively abandoned in case it should discourage investment. The incentives were widely available: by 1979 the assisted area map covered 44 per cent of the working population. The changes introduced in 1979 were a reduction in the assisted areas to 26 per cent of the working population, and a cut in level of Regional Development Grant in DAs and IAs. The cut in coverage meant that assisted areas became fragmented rather than compact. This was expected to produce a 40 per cent cut in spending within four years – but in the event produced only a 10 per cent cut. A slight increase in assisted area coverage was made in 1982, but major changes in regional policy were announced only in 1984, following a review of the policy. This widened the coverage of assisted areas by including much of the West Midlands, but eliminated SDAs. It sought to shift the balance of incentives towards selective assistance. RDG would now be subject to a cost per job limit and would not be given for replacement investment, but certain services industries would be eligible. Again, a large fall in spending was expected – but by 1988 had not materialized (HCP 346).

The 1979 and 1984 changes reflect a balancing of a variety of interests. Advocates of a complete abandonment of regional policy (on the right) and an expansion (on the left) were ignored. The more influential voices were those favouring greater cost-effectiveness. Critics of the policy's manufacturing focus, high cost per job and unselectiveness were listened to.[11] And regional interest groups were strong enough to ensure that advocates of drastic cuts in spending on the policy did not get their way. A third influence was the government's strategy vis-à-vis the EEC. Regional aid from the EEC is only available for projects in areas that are designated as assisted areas for national regional policy, and the European Regional Development Fund (ERDF) has set a 35 per cent limit to the coverage of the population by assisted areas. This explains the British government's willingness to raise the assisted area coverage to 35 per cent from 28 per cent.

The rationale for regional policy has also undergone some important changes. From 1975, under the Labour government, regional policy became a means of industrial modernization: incentives were available in the regions even to firms that were installing machinery that would lead to an overall reduction in their labour force. The argument was that otherwise such plants might cease to exist at all. In addition, the attraction and retention of internationally mobile firms became an even more significant aim of policy. But by 1975 two key assumptions of regional policy were starting to be undermined: the belief that large regions *as wholes* had common problems and the 'growth pole' rationale for regional policy.

The attack on 'broad' regions was a reflection of new patterns of restructuring, which were magnifying intra-regional differences and producing new spatial patterns of advantage and disadvantage. A widely quoted finding was the growth of employment and residential population in small town and semi-rural environments outside big cities (Fothergill and Gudgin, 1982). At the policy level, however, it manifested itself through a re-evaluation of inner city policy.

The failure at anything but a symbolic level of the social and environmental-based policies of the 1966–75 period led to a new direction in inner city policy.

This had two features. There was a continued emphasis on cooperation with local authorities; and there was an acknowledgement of job loss as a force shaping opportunities in inner cities – this was indeed a conclusion of the research carried out under the previous phase of inner city policy. In 1977 'partnerships' were set up between central government and seven authorities, and half of the expanded Urban Programme funds were allocated to them. Following lobbying by excluded authorities, two tiers of thirty 'programme authorities' and 'other districts' were created where problems were less pronounced. The result was a moderate inflow of resources for capital investment mostly, including industrial area improvement and other purposes related to economic development. These had to be supported on a 25 per cent basis by councils' own funds. Again, the policy has a strong symbolic aspect as the resources were limited (in the £ hundreds m.). More important was the Labour government's attempt to redirect resources to inner city authorities through the rate support grant (Bennett, 1982). But the fact that Birmingham and London had inner city problems, despite being in affluent regions, weakened support for the idea that regions were either uniformly affluent or uniformly depressed. The Labour government did not go so far as to abandon its regional policy in consequence of this. However, it did announce that no further New Towns would be built and that expansion plans in existing New Towns would be scaled down. (Symbolically, in 1977, the plan for Stonehouse New Town near Glasgow was dropped and a large injection of funds into Glasgow inner areas was announced instead.) The grounds for this policy change were that by providing housing and jobs outside existing centres, New Towns policy was a cause of job loss in the inner city. This argument was dubious, as the scale of New Town expansion was relatively small and most of the firms that set up there would probably not have done so in inner cities.

It was only in 1979, with the return of a Conservative government, that the implications of this more differentiated concept of regional economic performance were drawn for regional policy. The cutback of the assisted areas and more selective targeting of incentives were justified as realistic responses to the differentiated economic situation within regions. They were also facilitated by the waning of belief in regional growth poles. Increasingly, it was argued, owing to greater concentration of corporate ownership and the insignificance of transport costs, factories were parts of large corporations with units spread throughout the country and had nationwide supplier and customer linkages, rather than linkages with local firms. Hence their capacity to produce multiplier effects in the regional economy is negligible and one justification for regional policy is removed.

Only in 1984 was a new justification of regional policy advanced: it was portrayed as a 'social policy' not really capable of shifting the spatial pattern of employment and hence not deserving of very much spending. This was something of a compromise. In terms of economic ideology the government's rejection of Keynesianism and commitment to improving national industrial competitiveness implied that the regional equity was not a policy goal. But the strength of vested interests prevented an abandonment of regional policy. The new 'social' rationale 'made sense' of the cuts in assisted area coverage and the intention of lower spending.[12]

The period since 1975 has seen a great expansion of the role of European assistance. In the period 1977–86, when British spending on regional incentives rose from £437m to £642m (a fall of 13 per cent in real terms), grants from the ERDF have risen from £95m to £363m (an increase of 80 per cent in real terms) (CSO, various). In particular areas, European assistance can be a substantial share of the total. Figures for the Clydeside conurbation show that between 1979 and 1983, UK regional incentives totalled £137m, and ERDF grants totalled £122m (Lever, 1986). However, most ERDF grants are given to local authorities for infrastructure assistance, rather than to firms.

Finally, there has been a considerable expansion in what Martin (1987) calls 'central government localism' in the post-1979 period. This consists of measures designed to bypass local authorities or to weaken them. Some of these were mentioned in Section 2 (Housing Action Trusts, etc.). The three with most relevance to economic development are Enterprise Zones, Urban Development Corporations and inner city policy.

The common theme of EZs and UDCs is that local government is 'part of the problem' and that economic activity is best fostered by facilitating market forces, which means reducing central and local government regulation. Both points mark a sharp break with previous policy.

The first two Urban Development Corporations were announced in 1981 (London Docklands, Merseyside) and they have been followed by nine more in 1987 and 1988 (Black Country, Trafford Park, Teesside, Tyne and Wear, Cardiff, Bristol, Leeds, Central Manchester and Sheffield). Their function is to develop land, infrastructure, housing and other facilities to promote economic development. Private capital is to be used where possible. They have certain planning powers, but remain subordinate to local authority plans (contrary to original intentions). Experience of the two first UDCs reveals the importance of demand: the London Docklands DC has been working in a far more favourable context than Merseyside DC, owing to restructuring in the financial services, printing and related industries. The absence of an elected element, however, has enabled the UDCs to give a low priority to social considerations.

Enterprise Zones are another approach to encouraging development that involves keeping local government at arms' length. Firms locating in EZs obtain two main benefits: a ten-year rates holiday, and 100 per cent tax allowances on industrial and commercial buildings. These advantages are to some extent mitigated by the higher rents charged in EZs: it is estimated that rents have increased 27 per cent on average, and that in one EZ this allowed 60 per cent of the benefits to be captured by owners of land and industrial property (Department of the Environment, 1987; Erickson and Syms, 1986). Eleven EZs were announced in 1980 and twelve more in 1982–3 (all but three in the Midlands, North, Scotland, Wales or Northern Ireland). Most of the first wave were in old industrial areas, but the second wave were more widely dispersed. They have taken over some of the functions of Special Development Areas in responding to major plant closures. An evaluation in 1987 showed that they had a net cost of £300m (1981–6) (made up of tax expenditure on capital allowance, grants to local councils in lieu of rates, etc.) and had created 12 860 net additional jobs in the local economy, i.e. a cost of £23 000 per job (Department of Environment, 1987).) Enterprise Zones were intended to show that freedom from state control can create jobs; but what they actually show is that public spending can create jobs. The original laissez-faire concept of an Enterprise Zone was not applied owing to the resistance of central departments and local authorities (Taylor, 1981).[13]

Inner city policy since 1979 has been marked by three main trends: a decline in the role of local authorities, a greater involvement of industrial and financial groups, and an increased dependence on private funds.

The Urban Programme inherited from Labour was continued and its funding increased, but local authorities have become subordinates rather than partners. This became particularly clear with the launching of five City Action Teams (in 1985) in partnership areas, made up of regional directors of the main government departments involved, whose role was to strengthen central influence over the choice of programmes.

There have also been other new initiatives. In 1982 the Business in the Community umbrella group of leading industrialists, financiers and civil servants and DoE regional offices was set up to help stimulate the creation of local Enterprise Agencies (Urry, 1987). In 1983 Urban Development Grant was launched to lever in private funds in Urban Programme Areas, and in 1986 Urban Regeneration Grant was announced to enable direct dealing between the Department of Environment and developers. The emphasis in all these initiatives is towards capital projects. Other initiatives include the Merseyside Task Force launched in 1981 and the sixteen 'mini task forces' set up in 1985 and 1987 in districts of cities. Their aim is to create jobs and training opportunities for local people and to encourage local enterprise (Stewart, 1987; Harding, 1988). In general, however, the level of funding is very low compared with the Urban Programme. But even the Urban Programme does little to make good the reductions in rate support grant in recent years. For example, Manchester, Birmingham and Liverpool respectively suffered block grant losses of 33, 20 and 15 per cent in real terms between 1981 and 1987 (Hegarty, 1988). This is due to the combined effects of socio-economic trends such as population loss and the reversal of the trend under Labour between 1974 and 1979, which diverted more grant to big city authorities.

Local economic policy. The period since 1975 has seen a further expansion of local economic development activity by local councils. In addition to previous legislation, in 1978 the Inner Urban Areas Act gave some fifty inner city local authorities new powers in the sphere of industrial assistance, cooperative formation, land development and industrial area improvement. And in 1982 local authorities gained new powers to assist private sector provision of industrial land and buildings.

The current level of such activity can be seen in Table 1.1, which shows the level of economic spending in £ per 1000 population for 1984. This implies that a county council with a population of 1 m will have an average budget of £2.3m, and a district council with a 100,000 population will have an average budget of £584,200. (These figures are

Table 1.1 *Economic development spending (in £ per 1000 population) of councils in England and Wales, for 1984.*[14]

	Metropolitan areas	Non-metropolitan areas	All areas	AAs	Non-AAs
County Councils					
Revenue (gross)	2258	600	1203		
Capital	1921	611	1115		
Total	4179	1211	2318		
District Councils					
Revenue (gross)	4283	2581	3172	4649	2670
Capital	3460	2235	2670	5055	1853
Total	7743	4816	5842	9704	4523

(*Source:* CIPFA figures for 1983–4 and 1984–5 averaged, quoted by Armstrong and Fildes, 1988, Tables 3 and 5).

not dissimilar to those quoted for 1984 by Mills and Young (1986), i.e. £1.5m per council.) For the nation as a whole they imply total spending on economic development of some £450m per year (2.318 + 5.842) × 55m). The proximity of this figure to the level of spending on regional policy in the form of Regional Development Grant and Regional Selective Assistance (£521m in 1983–4) shows the significance of local council spending. The table also reveals the extent of economic development activity, as does the Mills and Young survey. It is not restricted to big city authorities, but includes small town and semi-rural areas. This means that the localities that have experienced the highest job loss are having to compete with areas that may have more 'natural' attractions for new firms. Economic development activity, however, does remain more pronounced in assisted areas than in non-assisted areas.[15]

Underlying this expansion are two forces. The first is the deteriorating unemployment situation. If the 1960s were the decade when regional unemployment again became a political issue, the period since the mid 1970s has been one of rising national unemployment (until 1987), combined latterly with growth. The lower level of investment from the mid 1970s and the weaker effect of regional policy (as the 'stick' element was removed and, especially after 1979, as the coverage of assisted areas was reduced) put more pressure on councils to expand their intervention. The second force behind this increased activity is the reaction by Labour councils to the Conservative general election victory of 1979. In response to the new government's claim that 'There is no alternative' certain Labour councils – notably the Greater London Council (from 1981), West Midlands County Council and Sheffield City Council – decided to demonstrate the opposite. The best known policies are probably the cheap transport fares policy of the GLC and South Yorkshire councils. But it was important to Labour to demonstrate that a national economic policy containing a strong planning element was viable. Since the 'Alternative Economic Strategy' elaborated in the 1979–81 period could not be implemented nationally, certain Labour councils decided to apply some of its ideas at local level, to demonstrate that they could work.

These initiatives started from a critique of previous forms of local economic intervention, such as the provision of land, premises and financial assistance to industry. It was argued that the provision of sites and premises is misdirected because they are not in short supply; that economic support is an unjustified subsidy to private enterprise; that job creation has been a subsidiary policy aim; and that, in aggregate, economic development policy does not create extra jobs but merely shifts jobs around[16] (Boddy, 1984; Mawson and Miller, 1986). In brief, traditional local economic action was not sufficiently interventive: it amounted to a subsidy to private capital without securing any change in its behaviour.

In contrast, the radical Labour councils sought to ensure restructuring 'for labour' rather than 'for capital'. To do so they adapted two features of the Labour party's national economic policy: 'planning agreements' and a National Enterprise Board. They sought to make planning agreements with the firms to which assistance was given, on the grounds that in exchange for assistance firms should accept some 'social' objectives in their investment plans and recruitment policies, and that work conditions and trade union rights should be protected. Secondly, they set up wholly controlled Enterprise Boards to manage their economic interventions, including the drawing up of planning agreements. One of the attractions of these Boards is that they can hold shares, whereas local councils can do so only under specific circumstances (CIPFA, 1988). Enterprise Boards were particularly attractive to county councils, whose resources are larger (McKean and Coulson, 1987); Sheffield, on the other hand, chose to work through a newly created Employment Department. Two further features appear in certain radical Labour council initiatives. One is a return to the Fabian notion of the council as model employer, and as a major actor in the local economy. This led to an emphasis on the council's own productive capacity (e.g. the design of 'socially useful' products) and its power to impose conditions on suppliers of goods and services via 'contract compliance'. The other is a commitment to popular democratic inputs into economic development policy. This was particularly strong at the GLC. Clearly the potential for tension between the top-down planning orientation of Enterprise Boards and the bottom-up orientation of popular planning institutions is considerable.

In these ways economic development policy was expected to make a break with previous policy: workers' needs would be placed above those of capital, 'hands-on' intervention would be practised, jobs in existing firms would be protected rather than new firms being attracted, and social criteria would be introduced into private sector investment and recruitment decisions. In addition to this new stance vis-à-vis the private sector, councils would make greater use of their own resources to demonstrate alternative patterns of economic development.

In reality there has been considerable diversity among radical Labour councils, and considerable change over time as some of the most ambitious proposals have proved unworkable. For example, planning agreements have rarely been imposed because the firms involved were economically weak owing to years of under-investment and/or threatened to refuse assistance if planning agreements were imposed.

As a result, Enterprise Boards have often ended up as sources of capital for firms that had exhausted other sources, and have not imposed social criteria. Indeed, the literature is more abundant on the hopes of radical council intervention than on what happened in practice. One evaluation is that the jobs protected or created by such policies as these have been relatively cheap – at £2500 to £4500 per job, which compares with £35000 per job under regional policy. But they have been relatively few in number: 8000 between 1981 and 1984 by the five main Enterprise Boards (Mawson and Miller, 1986, p. 170).

Critics, such as Eisenschitz and North (1986), would argue that the idea of 'restructuring for labour' is ill-thought-out and that, in practice, Enterprise Boards have mainly increased the rationality of capitalism by helping the restructuring of small firms, which would otherwise not have taken place. Secondly, they argue that market constraints are severe, e.g. once 'socially useful' products have been designed they still have to find capitalist producers who judge them profitable to make; also, workers' cooperatives may impose low wages and poor conditions in order to survive. Thirdly, they point out that estimates of job creation are often exaggerated, because they ignore dynamic competitive effects such as job losses in less competitive firms.

Whatever the economic and social effects of the economic development policies of radical Labour councils, the councils themselves would claim that one of their prime aims has been political – to demonstrate within their locality that alternative policies were possible – and in this respect they have undoubtedly been successful. Indeed, the success of the GLC and metropolitan county councils helped lead to their abolition.

In brief, for both economic and political reasons local economic development policy has expanded greatly since the mid-1970s. This is indeed an international trend. At the same time, the type of policy adopted – from the 'traditional' to the 'radical' – has been variable between councils. This reinforces our earlier argument that permissive legislation allows councils of different characters to develop policies of different kinds. The key feature of the 'radical' Labour councils (which are a minority of all Labour councils) is the distinctiveness of the councillors involved. As Gyford (1985) has argued, they are drawn from a 'new urban left' made up of highly educated often public sector employees in the service sector. This is in part a generational phenomenon: the new councillors were often involved in the community action movement of the 1960s, in housing cooperatives, etc., and are committed to an alternative, decentralist form of local government. By comparison, where the Labour group is dominated by manual trade unions, as in Liverpool or Dundee, policies are more traditional (Elliott and McCrone, 1984). This difference is particularly clear in Sheffield, where the Labour group changed composition in 1980 from an older 'right wing' group with steel and engineering union connections to 'young left-wing councillors with professional jobs' (Duncan and Goodwin, 1988, p. 84) with rapid consequences for policy. We shall return to the question of the explanation of the diversity of policy between councils in the Conclusion of the book (Chapter 9).

(v) CONCLUSION

In sum, public intervention in the economic fortunes of localities is on a vastly greater scale than ever before. The continuation of regional policy by a government ideologically opposed to it indicates the strength of the beneficiary interests. In particular, it is an important means of responding to regional and sub-regional interest groupings (Pickvance, 1985). Local council economic activity is now on the same scale as regional policy, and this is a new development. Government response, however symbolic, to local economic problems is also shown by inner city policy, Enterprise Zones, etc. Last but not least there is an ever-increasing array of European and other schemes by which local areas can attract resources. For example, the European Regional Development Fund, European Social Fund, European Coal and Steel Community, European Investment Bank, (English) Development Commission, and English Tourist Board, all offer area-based as well as generally available schemes providing grants and cheap loans.

However, this places an ever-greater premium on local authority knowledge and capacity to pursue resources. This involves two phases. The first is influence on the creation of schemes or acquisition of powers. The existence of special status areas, such as assisted areas, or ECSC Priority Employment Areas, and their precise boundaries, is not given from above. Undoubtedly they reflect economic rationales, but the number and nature of the criteria used to define them are open to argument, and local interests can hope to influence these criteria. For example, 'peripherality' is one of the many criteria of assisted area designation, but probably owes as much to Scottish and Welsh lobbying as to pure economic considerations. Likewise, the powers available to local authorities for economic development in part reflects local initiative. The Inner Urban Areas Act of 1978 among other things made generally available powers obtained by Tyne and Wear County Council under a private Act in 1976 (Rogers and Smith, 1977).

The second phase where local authority dynamism or lack of it makes a difference is in the acquisition of special area statuses, or the implementation of available powers. Some special statuses are decided by central government or the EEC; others require an application to be made. In the former case the consent of local authorities may or may not be a condition. (In the early days of Enterprise Zones some local councils resisted them (Taylor, 1981).) However, even schemes that appear to involve the application of clear-cut criteria may depend on initiatives from local areas. Assisted area designation involves a specific set of criteria (Cmnd 9111). However, they are so numerous that it is worthwhile for many councils to submit a case for inclusion (Townsend, 1980; Pickvance, 1985). The role of demand is also well demonstrated in the case of the West Midlands, which for many years resisted the idea of assisted area status and hence delayed its designation (Pickvance, 1986b; Morgan, 1985).

The implementation of local authority powers also depends on councils' assessments (which depend in part on legal opinions) of the extent of these powers. For example, local councils do not have an explicit power to own equity, but have used section 137 to justify this. Some councils will hold back because of the lack of explicit powers while others will go ahead regardless of consequences. Again, we are brought back to the question of the dynamism of particular councils.

The aim of this introduction has been to outline the institutional context affecting the economic fortunes of localities. We hope to have shown that these fortunes depend as never before on councils' capacity to negotiate their way around a new political economy in which private financiers, industrialists, and European institutions are as important as central government. The chapters that follow will show the extent to which the seven localities have been favoured by 'market forces' and central government policies, and the types of economic initiative that the councils in each area have taken. In Chapter 9 (Conclusion) we shall seek to explain the patterns observed.

Notes: Chapter 1

1. For example, critics have pointed out that the expected patterns do not occur in every country. But, as Saunders points out, he is discussing a tendency and does not expect empirical universality.
2. See also Miller (1986). A recent study shows that rates levels *do* influence voting shares at local elections but that this does not often affect the number of seats held (Gibson, 1988). This explains how council control may remain uninfluenced by local policy even when it influences voting patterns.
3. Since many welfare state citizenship rights are a matter of central policy, and local government capacity to affect private property rights (in land and housing) is striking, it is peculiar that Saunders links citizenship and private property ideologies with local and central government respectively.

4 Detailed accounts of central government policy towards local government are available elsewhere, so here we shall focus on the main points only: Loughlin (1986), Pickvance (1986a), Rhodes (1988), Duncan and Goodwin (1988).

5 The deflator used in this and similar calculations is the general government expenditure deflator.

6 By contrast, the 1981 controls on capital spending (which replaced the previous controls on borrowing) had more success. They led to a very sharp decrease in capital spending, but gave rise to a new problem of unpredictability. 'Underspending' between 1981 and 1983 was replaced by 'overspending' between 1983 and 1985. Unpredictability is partly inherent in capital projects that are 'lumpy' and where spending is spread out over a long period. Underspending was also due to the fact that capital projects entail current spending (on wages, supplies) and councils feared this would lead to penalties, and 'overspending' to the build-up of revenues from council house sales.

7 There are serious problems in measuring financial squeeze. Central grants rose 0.3 per cent per year faster than inflation during the 1979–87 period (CSO, 1988), but to meet needs at a constant level a faster rate of increase would have been necessary – quite how fast depends on the hard-to-pin-down concept of 'need'.

8 By 1990, councils will have to introduce tendering in refuse collection, cleaning (of buildings and streets), catering, ground maintenance and vehicle repair and maintenance. In sports and leisure it will be introduced later (Department of Environment, 1988). Note that in all of these spheres manual workers are mostly involved. However, the Minister has discretion to add to this list of services.

9 Seven levels of public intervention have been distinguished by Urry (1987): European (EEC), central government (regional policy), regional (Scottish Development Agencies, English Industrial Estates Corporation), nationally organized urban policies (Enterprise Zones, etc.), county council, district council and community level initiatives. We shall be mainly concerned with regional policy and council initiatives, though the other types will be mentioned.

10 Rates are a tax levied on the rateable value of buildings in an area. (Rateable values depend on rent levels, but are adjusted only infrequently.) A council where total rateable value is £10m will be able to obtain £15m in rates by levying a rate of £1.50 per £ of rateable value. The product of a 1d rate (or 2p as it later became) is thus the yield from applying a 1d (or 2p) rate. It is bigger for county councils than for district councils owing to their larger rateable bases.

11 In 1988 a further policy change was announced: RDG would cease to be available. This can be seen as a belated response to criticism of the deadweight element in it, i.e. the fact that in many cases RDG did not change firm investment behaviour, but took the form of a windfall, cutting the cost of decisions firms were making anyway.

12 Too much importance should not be given to rationales for policies. Ideas are not sole or even prime movers of policies. Equally, as the above

example shows, rationales for policy are means by which governments can give their policies a coherence and plausibility they would otherwise lack. Nevertheless, even if policies change owing to the balance of power between different political groupings – in this case segments of the Conservative party, government departments, regional pressure groups among others – ideas are an element in this balance of power.

13 For completeness we should mention two other schemes. Six Free Ports were launched in 1984 in Birmingham, Cardiff, Liverpool, Southampton, Belfast and Prestwick, allowing freedom from duties and related paperwork. They appear to have attracted little interest. Simplified Planning Zones were introduced in 1986 as a form of Enterprise Zone without incentives, but there is no information on their experience.

14 In addition, there has been an increase in the level of guarantees given to firms. This only shows up in spending data if a firm cannot repay a loan (CIPFA, 1988).

15 The change in the rating system will remove one of the ideological supports for local economic development activity. Proponents of local growth have always emphasized its alleged effect of increasing the local rateable base (and hence reducing rates levels) as well as increasing employment, retail turnover, etc. (In fact, the equalization element in central grant meant that increases in rateable value were compensated by lower grant.) But under the 1990 system, the fact that businesses pay a centrally collected uniform business rate whose proceeds are then distributed among councils on a per capita basis will make this belief less credible, because increases in non-domestic rateable value will have no direct impact on local fiscal resources. Other arguments for growth, such as increased jobs and increased retail turnover, however, will remain tenable.

16 It is interesting that this self-defeating process was an element in the government's attempt in 1982 to cut the 'free 2p' rate to ½p – a proposal that was successfully resisted by the local authority associations (Mawson and Miller, 1986).

References: Chapter 1

Aldridge, M. 1979. *The British new towns.* London: Routledge and Kegan Paul.

Armstrong, H. and Fildes, S. 1988. 'District council industrial development initiatives and regional industrial policy', *Progress in Planning*, **30**, 85–156.

Ascher, K. 1987. *The politics of privatisation.* London: Macmillan.

Ashford, D. E. 1980. 'Central–local financial exchange in the welfare state'. In D. E. Ashford (ed.), *Financing urban government in the Welfare State.* London: Croom Helm.

Ashford, D. E. 1982. *British dogmatism and French pragmatism.* London: Allen and Unwin.

Bennett, R. J. 1982. *Central grants to local authorities.* Cambridge: Cambridge University Press.

Berthoud, R. 1976. 'Where are London's poor?', *GLC Intelligence Quarterly*, **36**, 5–12.
Boddy, M. 1984. 'Local economic and employment strategies'. In M. Boddy and C. Fudge (eds), *Local Socialism?* London: Macmillan.
Briggs, A. 1968. *Victorian cities*. Harmondsworth: Penguin.
Buck, N. H. 1981. 'The analysis of state intervention in nineteenth-century cities: the case of municipal labour policy in east London 1886–1914'. In M. Dear and A. J. Scott (eds), *Urbanization and urban planning in capitalist society*. London: Methuen.
Bulpitt, J. G. 1983. *Territory and power in the United Kingdom*. Manchester: Manchester University Press.
CIPFA. 1988. *Financial Information Service: Vol. 31 Financial resources for economic development*. London: Chartered Institute of Public Finance and Accountancy.
Camina, M. M. 1974. 'Local authorities and the attraction of industry', *Progress in Planning*, **3** (2), 83–182.
Cawson, A. and Saunders, P. 1983. 'Corporatism, competitive politics and class struggle'. In R. King (ed.), *Capital and politics*. London: Routledge and Kegan Paul.
Central Statistical Office. 1988. *U.K. National Accounts*. London: HMSO.
Cockburn, C. 1977. *The local state*. London: Pluto.
Cmnd 6153, 1940. *Royal Commission on the distribution of the industrial population* (Barlow Report). London: HMSO.
Cmnd 4040, 1969. *Royal Commission on local government in England 1966–9* (Maud Report). London: HMSO.
Cmnd 6453, 1976. *Report of the Committee of Enquiry. Local government finance* (Layfield Report). London: HMSO.
Cmnd 9111, 1983. *Regional industrial development*. London: HMSO.
Crispin, A. 1983. Comment on 'The case of education'. In K. Young (ed.), *National interests and local government*. London: Heinemann.
Department of the Environment. 1987. *An evaluation of the Enterprise Zone experiment*. London: HMSO.
Department of the Environment, 1988. *Local Government Act 1988: Part I and Schedule I – Competition in the provision of local authority services* (Circular 19/88). London: HMSO.
Duncan, S. S. and Goodwin, M. 1988. *The local state and uneven development*. Cambridge: Polity.
Dunleavy, P. 1979. *Urban political analysis*. London: Macmillan.
Eisenschitz, A. and North, D. 1986. 'The London industrial strategy: socialist transformation or modernizing capitalism?', *International Journal of Urban and Regional Research*, **10**, 419–40.
Elkin, S. L. 1974. *Politics and land use planning*. Cambridge: Cambridge University Press.
Elliott, B. and McCrone, D. 1984. 'Austerity and the politics of resistance'. In I. Szelenyi (ed.), *Cities in recession*. London: Sage.
Erickson, R. A. and Syms, P. M. 1986. 'The effects of enterprise zones on local property markets, *Regional Studies*, **20**, 1–14.
Eyles, J. 1979. 'Area-based policies for the inner city: context, problems and

prospects'. In D. T. Herbert and D. M. Smith (eds), *Social problems and the city*. Oxford: Oxford University Press.

Forrest, R. and Murie, A. 1985. 'Restructuring the welfare state: the privatisation of public housing in Britain'. In W. van Vliet, E. Huttman and S. Fava (eds). *Housing needs and policy approaches: international perspectives*. Durham NC: Duke University Press.

Fothergill, S. and Gudgin, G. 1982. *Unequal growth*. London: Heinemann.

Friedland, R., Piven, F. F. and Alford, R. R. 1977. 'Political conflict, urban structure and the fiscal crisis', *International Journal of Urban and Regional Research*, **1**, 447–71.

Gibson, J. G. 1988. 'Rate increases and local elections: a different approach and a different conclusion', *Policy and Politics*, **16**, 197–208.

Gyford, J. 1985. *The politics of local Socialism*. London: Allen and Unwin.

Hamnett, C. 1979. 'Area-based explanations: a critical appraisal'. In D. T. Herbert and D. M. Smith (eds), *Social problems and the city*. Oxford: Oxford University Press.

Harding, A. 1988. 'Spatially specific urban economic development programmes in Britain since 1979', *West European Politics*, **11** (1), 102–15.

Harrington, T. 1984. *Explaining local authority policy-making: class pressures, professional interests and systems of political management in two county councils*, PhD thesis, University of Kent at Canterbury.

HCP 346, 1987. 'Report of Comptroller and Auditor General', *Regional Industrial Incentives* (1987–88 session). London: HMSO.

Hegarty, S. 1988. 'Inner city aid: the figures behind the hype', *Public Finance and Accountancy*, 8 January 1988, 7–10.

Keeble, D. 1976. *Industrial location and planning in the UK*. London: Methuen.

Kellett, J. R. 1978. 'Municipal socialism, enterprise and trading in the Victorian city', *Urban History Yearbook 1978*, 36–45.

King, D. S. 1989. 'Political centralization and state interests in Britain: the 1986 abolition of the GLC and MCCs', *Comparative Political Studies*, **21**, 467–94.

Laffin, M. 1986. *Professionalism and policy: the role of the professions in the central–local government relationship*. Aldershot: Gower.

Lawless, P. 1986. *The evolution of spatial policy*. London: Pion.

Leach, S., Game, C., Gyford, J. and Midwinter, A. 1986. 'The political organisation of local authorities'. In Cmnd 9798 (Widdicombe Report). *Committee of Inquiry into the conduct of local authority business*, Research Volume I. London: HMSO.

Lever, W. 1986. 'Old policies in a new role'. In W. Lever and C. Moore (eds), *The city in transition*. Oxford: Oxford University Press.

LEWRG, 1979. *In and against the state*. London: London Edinburgh Weekend Return Group.

Loughlin, M. 1986. *Local government in the modern state*. London: Sweet and Maxwell.

McCrone, G. 1969. *Regional policy in Britain*. London: Allen and Unwin.

McKean, B. and Coulson, A. 1987. 'Enterprise Boards and some issues raised by taking equity and loan stock in major companies', *Regional Studies*, **21**, 373–84.

Martin, R. L. 1985. 'Monetarism masquerading as regional policy? The government's new system of regional aid', *Regional Studies*, **19**, 379–88.

Martin, R. L. 1988. 'The new economics and politics of regional restructuring: the British experience', in L. Albrechts *et al.* (eds), *Regional policy at the crossroads*. London: J. Kingsley.
Martin, R. L. and Hodge, J. S. C. 1983. 'The reconstruction of British regional policy' (two parts), *Government and Policy*, **1**, 133–52, 317–40.
Mawson, J. and Miller, D. 1986. 'Interventionist approaches in local employment and economic development: the experience of Labour local authorities'. In V. Hausner (ed.), *Critical issues in urban economic development*, Vol. 1. Oxford: Clarendon Press.
Miller, W. L. 1986. 'Local electoral behaviour'. In Cmnd 9800 (Widdicombe Report). *Committee of Inquiry into the conduct of local authority business*, Research Volume III. London: HMSO.
Mills, L. and Young, K. 1986. 'Local authorities and economic development: a preliminary analysis'. In V. Hausner (ed.), *Critical issues in urban economic development*, Vol. 1. Oxford: Clarendon Press.
Morgan, K. 1985. 'Regional regeneration in Britain – the territorial imperative and the Conservative state', *Political Studies*, **33**, 560–77.
O'Connor, J. 1973. *The fiscal crisis of the state*. New York: St Martins Press.
Page, E. and Goldsmith, M. 1987. 'Centre and locality: explaining cross-national variation'. In E. Page and M. Goldsmith (eds), *Central and local government relations: a comparative analysis of West European unitary states*. London: Sage.
Pahl, R. E. 1975. *Whose city?* (2nd edn). Harmondsworth: Penguin.
Parkinson, M. 1986. 'Creative accountancy and financial ingenuity in local government: the case of Liverpool', *Public Money*, March 1986, 27–32.
Parsons, D. W. 1986. *The political economy of British regional policy*. London: Croom Helm.
Pickvance, C. G. 1981. 'Policies as chameleons: an interpretation of regional policy and office policy in Britain'. In M. Dear and A. J. Scott (eds), *Urbanization and urban planning in capitalist society*. London: Methuen.
Pickvance, C. G. 1985. 'Spatial policy as territorial politics: the role of spatial coalitions in the articulation of "spatial" interests and in the demand for spatial policy'. In G. Rees *et al.* (eds), *Political action and social identity*. London: Macmillan.
Pickvance, C. G. 1986a. 'The crisis of local government in Great Britain: an interpretation'. In M. Gottdiener (ed.), *Cities in stress*. Beverley Hills, Calif.: Sage.
Pickvance, C. G. 1986b. 'Regional policy as social policy: a new direction in British regional policy?'. In M. Brenton and C. Ungerson (eds), *Yearbook of social policy in Britain 1985–6*. London: Routledge and Kegan Paul.
Pinch, S. 1985. *Cities and services: the geography of collective consumption*. London: Routledge and Kegan Paul.
Rhodes, R. A. W. 1988. *Beyond Westminster and Whitehall*. London: Unwin Hyman.
Rogers, P. B. and Smith, C. R. 1977. 'The local authority's role in economic development: the Tyne and Wear Act 1976', *Regional Studies*, **11**, 153–63.
Saunders, P. 1983. 'The regional state: a review of the literature and agenda for research', *University of Sussex Urban and Regional Studies Working Paper No. 35*, Brighton: University of Sussex.

Saunders, P. 1986. 'Reflections on the dual politics thesis: the argument, its origins and its critics'. In M. Goldsmith and S. Villadsen (eds), *Urban political theory and the management of fiscal stress*. Aldershot: Gower.

Sharpe, L. J. 1979. 'Modernising the localities: local government in Britain and some comparisons with France'. In J. Lagroye and V. Wright (eds), *Local government in Britain and France*. London: Allen and Unwin.

Sharpe, L. J. 1985. 'Central coordination and the policy network', *Political Studies*, **33**, 361–81.

Sharpe, L. J. and Newton, K. 1984. *Does politics matter? The determinants of public policy*. Oxford: Oxford University Press.

Stewart, M. 1987. 'Ten years of inner city policy', *Town Planning Review*, **58**, 129–45.

Stoker, G. 1988. *The politics of local government*. London: Macmillan.

Taylor, S. 1981. 'The politics of enterprise zones', *Public Administration*, **59**, 421–39.

Townsend, A. R. 1980. 'Unemployment, geography and the new government's "regional" aid', *Area*, **12**, 9–18.

Urry, J. 1987. 'Economic planning and policy in the Lancaster district', *Lancaster Regionalism Working Paper No. 21*. Lancaster: Lancaster Regionalism Group.

2

Swindon: the rise and decline of a growth coalition

KEITH BASSETT and MICHAEL HARLOE

1 Introduction

Swindon's history has been one of remarkable growth and change. In 1841, when Brunel chose the town as the site for the Great Western Railway's locomotive works, the population was less than 2500. By 1911 it was 51 000 and the railway works, employing more than 10 000, dominated the local economy and shaped a distinctive, geographically isolated, working-class community set in the heart of rural Wiltshire. Railway employment began to decline in the 1930s and the works finally closed in 1986. But Swindon did not suffer the fate of many similar one-industry towns. It experienced rapid growth from the early 1950s and a transformation of its economic and social structure. By the 1980s the town had a population of 186 000 and a diversified economy based on new and high tech industries, offices, warehousing and distribution centres. It was being hailed in the press as 'the fastest growing micro-economy in Europe' and 'a high tech boom town'. All this, it was claimed, had been accomplished by a progressive, go-ahead council without any government assistance, representing (in the words of the town's chief executive) 'a model of urban development that is under local democratic control' (*Financial Times*, 9 November 1984).

In this chapter we shall analyse the council's prominent role in development and explain why it is now in decline. To do this, local developments have to be set in the wider context of changes in the South-East region as a whole and in state policies, which have provided Swindon with both constraints and opportunities. In responding to these opportunities the council has been able to draw upon a variety of resources, which may be classified as physical (land and environment), locational, economic (labour supplies and infrastructure), financial, political (powers of intervention and mobilization) and administrative. Constraints, opportunities and resources have all varied over time and can best be explored by linking policy shifts to successive rounds of

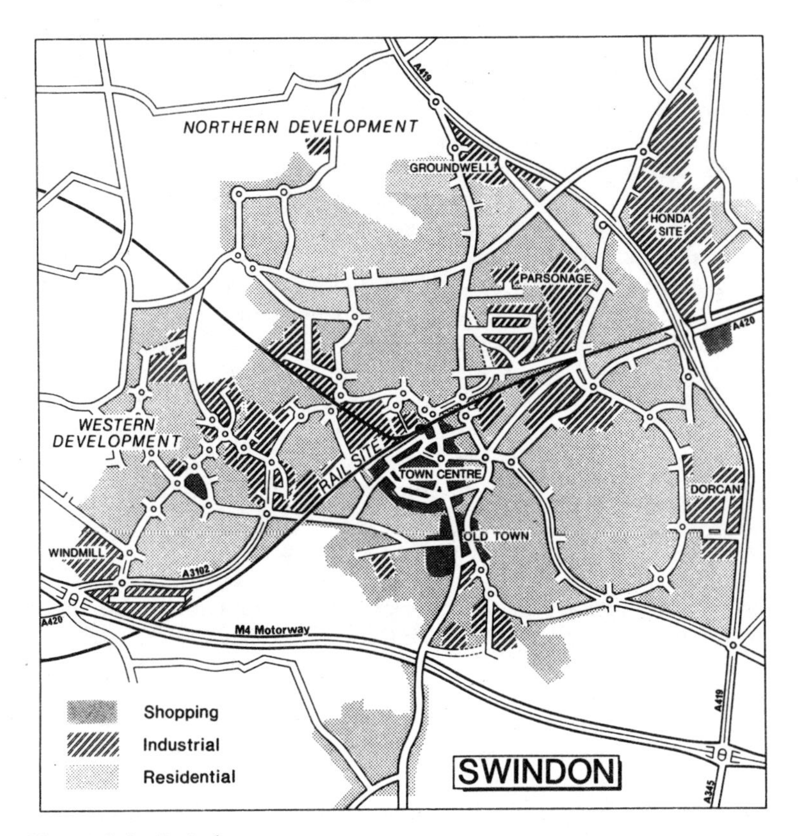

Figure 2.1 Swindon.

investment, economic restructuring, and social and political change. In what follows we distinguish a number of phases of development and, although we concentrate more on the latest period, Swindon today cannot be understood without the background of the earlier phases.

2 Reconstructing the locality, 1945 to the early 1980s

Prelude to expansion, from the 1930s to 1952: the emergence of a growth coalition

In 1939 the Swindon economy was still highly dependent on the rail workshops, although during the war some diversification resulted from the relocation of important defence plants to the town. The dangers of dependence on a single, major industry had been recognized by the council in the 1930s, and when Labour took control for the first time in 1945 it was able to secure general support for a policy of major expansion and diversification.

The term 'growth coalition' is a useful one to describe the combination of forces behind expansion. This coalition was led by a council, 61 per cent of whose members were railwaymen with close ties to the trades unions and deep roots in what was still a traditional occupational community with few evident class divisions. The coalition embraced the opposition Independents, representing the small petit bourgeois class of shopkeepers and traders. Leading officers in the council administration were strongly in favour, not least because of the enhanced powers and status that would flow from expansion. The dominant employer, the Great Western Railway, whose power had been weakened by nationalization, acquiesced in growth, and the new defence employers such as Plessey and Vickers were in favour in order to overcome labour shortages. In 1947 Swindon put forward its first long-term development plan, aiming for a more balanced community of 80–100 000.

The construction of a growth coalition was not enough, however, for the external policy context did not yet favour the realization of Swindon's goals. Although Swindon was able to retain many of its wartime industries, the Board of Trade sought to divert new investment to the newly designated Development Areas through the use of Industrial Development Certificates. The other major regional policy initiative of the period, the New Towns programme, also offered little hope. Swindon was regarded as being too big and too far from London to form the nucleus of a New Town. Progress had to await a change in government and a shift in the national policy context.

The first wave of expansion, 1952 to the late 1960s: Swindon and the Town Development Act

The Conservative Government's 1952 Town Development Act provided Swindon with its opportunity. It was intended to provide more rapid relief for the congested conurbations, at lower cost than building new towns, by subsidizing the expansion of existing small towns through overspill agreements. Swindon began negotiations with Tottenham to rehouse London families even before the Bill was passed and, with the reluctant acquiescence of Wiltshire County Council, secured ministerial consent for a population target of 92 000 after twenty years.

All those who have commented on Swindon's enthusiasm for town expansion in this period have pointed to the key role played by David

Murray John, the town clerk. Contemporaries attested to his skills as a negotiator and even described him as the 'prime mover' behind town expansion. However, the town clerk's prominent role should be seen in the broader political context. Expansion never became an issue that divided the parties; nor were there significant sources of opposition outside the council. When, in 1954, a prospective Conservative parliamentary candidate attacked the lack of publicity surrounding the expansion scheme the local paper proclaimed that 'expansion is not a political matter and we deplore the intrusion of politics into it' (quoted in Harloe, 1975, p. 230).

This low level of opposition permitted the concentration of power and decision-making in the hands of a few leading officers and councillors. This tight consensual arrangement provided an ideal context for initiating and pursuing a rapid expansion scheme, enabling the council to be run on lines somewhat akin to a development corporation. Although complaints that Swindon was being run by a benevolent but secretive and narrow-minded 'councilocracy' began to emerge in the 1960s, these views were those of a small, middle-class minority.

Under the Town Expansion scheme Swindon's population expanded by almost 30 000 (43 per cent) between 1951 and 1966. Growth was basically dependent on the attraction of new and mobile industry. However, although employment expanded rapidly the local economy did not become as diversified as was originally hoped. New firms such as Pressed Steel, which built a large car body plant in the town, were concentrated in the engineering sector, attracted by Swindon's reputation as a centre of skilled engineering labour. As a result, Swindon was still an overwhelmingly working-class town in the late 1960s, albeit with a preponderance of skilled rather than unskilled workers.

In the early years of expansion it was the promise of a job *and* a house that encouraged many families to migrate. Apart from a sprinkling of managers and professionals who preferred to live in the surrounding countryside, most first generation migrants became council tenants. At first, new development was not expensive. Central government paid special housing subsidies and the council was able to purchase compulsorily land at agricultural values. Later, when profits and rates from commercial and industrial development started to roll in, the council was able to continue its programme of land purchase.

By the late 1950s reasonably well-paid and secure employment and growing opportunities for women to work made home ownership increasingly attractive. The council responded by changing its development role. By using statutory powers and with some astute open market purchases the council came to own or effectively control access to most of the development land. Instead of building on this land itself, it leased it to private developers or formed partnerships with them. In this way the council exercised far more control over private development than planning powers alone would have allowed. At the same time financial benefits came to the local authority as land prices and rents escalated. This entrepreneurial approach increasingly dominated local authority policies and, by the late 1960s, the council had built up a highly professional and effective development organization. Essentially, Swindon was emulating the new towns, developing housing and industrial estates and appropriating developers' profits for the public sector.

The central shopping area was also transformed in the 1960s through the start of a major redevelopment programme, initially in partnership with a private developer but later financed by the local authority itself.

From about 1963 onwards, however, population and employment growth rates began to slow. The flow of migrants from London began to decline, and the Board of Trade began to take a tougher line in granting IDCs outside the Development Areas. It seemed as if Swindon's expansionary drive was running out of steam, when a shift in the national policy context opened another window of opportunity.

Relaunching expansion, 1968 to the early 1970s: Swindon as a 'new city' counter-magnet

Swindon was launched on a new wave of expansion on the basis of two developments. First, Swindon came to figure prominently as a 'New City' counter-magnet in regional plans for the South-East. The South East Study, published in 1964, was one of a series of regional plans that emerged in the 1960s. It recommended that one-third of the expected population growth in the region up to 1981 should be accommodated in new and expanded towns. Newbury–Hungerford was suggested as a possible site for a new city and Swindon as a site for major expansion. However, a follow-up consultant's report rejected expansion at Newbury in favour of Swindon. As Levin (1976) makes clear, the consultants were swayed by opposition in Berkshire and the obvious eagerness and proven track-record of Swindon. A further joint study by Swindon, Wiltshire and the GLC (known locally as 'The Silver Report') recommended a 1986 population target of 241 000, although the Secretary of State subsequently gave approval for a reduced figure of 200 000. Nevertheless, this still implied a commitment to a further massive expansion programme and an important role for Swindon in the expanding economy of the South-East region.

The second policy development that favoured Swindon was the decision to route the M4 motorway past the southern edge of the town.

When it was opened in 1971 it confirmed the town's nodal position on a high-speed transport corridor running westwards from London, past Heathrow, to Bristol and South Wales.

Locally, the political commitment to growth survived a change of council control. In 1968 the Conservatives, who had displaced the Independents as the chief opposition, took control and held power during the crucial period of relaunching expansion up to 1971. Although members of that administration have claimed that they began to reorient economic policy more towards an emphasis on electronics and office employment, the major thrust of development policy remained the same. One of their major initiatives was to centralize and streamline the administrative structure that coordinated this development, appointing a chief executive and consolidating departments into a small number of directorates. This structure remained largely unchanged into the 1980s.

Decentralization and the emerging M4 corridor: from the early 1970s to the mid-1980s

During this period the national policy of encouraging population to decentralize from the large cities to new and expanded towns was largely abandoned in the face of inner city decline. By this time, however, Swindon's growth was no longer dependent on overspill agreements with inner London boroughs. Strategically located on the M4 corridor, Swindon was in a strong position to take advantage of the reorganization of the regional economy of the South-East by attracting manufacturing plants and offices decentralizing from London.

In the early 1970s, encouraged by the local authority, new employers were already moving major plants and headquarters to the town, many from cramped and expensive sites in London. They included manufacturing firms such as Roussel, Raychem and Emerson Electrics, in growth sectors such as chemicals, pharmaceuticals and electronics, and service sector firms such as Hambro Insurance (now Allied Dunbar) and the Nationwide Building Society.

However, Swindon was not immune from the impact of industrial decline. Traditional employers such as the British Rail Workshops, Plessey (electrical engineering) and Garrards (record players) experienced major redundancies in the 1970s, pushing local unemployment rates temporarily above the national average between 1974 and 1976, and prompting press headlines like 'Boom town that ran out of steam' (*Western Daily Press*, July 1975). Indeed, local concern was such that in 1975, and again in 1978, the Council lobbied (unsuccessfully) for Intermediate Area Status.

Although the rise in unemployment rates proved temporary it helped to precipitate a reshaping and enhancement of the town's marketing strategy. The 1976 Corporate Plan had committed the council to the ambitious target of creating 3000 new jobs a year. In order to achieve this it was proposed to launch a more vigorous marketing strategy, build up a wider portfolio of sites, and appoint a new marketing manager. In 1976 the Conservatives briefly took control of the council again and were thus in a position to appoint the new marketing manager. The leadership agreed they needed someone with a business and marketing background to sell the town and meet industrialists 'over a gin and tonic in the City' (interview with authors, 1987). The post went to Douglas Smith, a former commercial director for Plessey with little sympathy for party politics.

Smith's new marketing unit, Swindon Enterprise, was granted a considerable degree of autonomy and, perhaps symbolically, was located not in the civic offices but at the top of the Murray John Tower, which dominated Swindon's townscape. Smith described his priorities as the attraction of high-tech companies and company headquarters. The advertising and marketing drive concentrated on Swindon's locational and environmental advantages. Smith's own version of the Swindon image was of a town '50 minutes by high-speed train to London, one hour Heathrow, next to the M4 in the golden corridor but cheaper than rival towns, surrounded by stunning countryside with lots of old rectories for executives to live in' (*Financial Times*, June 1986). Initial emphasis was placed on contacting US electronics and pharmaceutical companies and building up contacts with London-based property agents and advisers.

A key element of the new strategy was the acquisition of a wider portfolio of sites to attract new companies, with particular emphasis on the construction of new 'campus sites' on the periphery of the town, offering standards appropriate to 'the second industrial revolution'. Private sector interest in industrial development had been growing since the early 1970s, and in 1981 the St Martin's Property Corporation (owned by the Kuwaiti government) and Taylor Woodrow began construction of major campus sites on the periphery of the town. The former development was on land purchased from the council, but in general, in spite of pressure from officers, Labour refused to agree to a general policy of freehold disposals of its land, insisting on 125-year leases except in exceptional circumstances. As land prices escalated, however, the council found itself increasingly priced out of the land market and unable to replenish its land banks.

New firms continued to flood into the town, even during the 1979–83 recession. Although traditional employers were again hard hit, manufacturing employment still grew by 3 per cent, compared with an 11 per cent drop nationally, and service sector employment grew by nearly 9 per cent. Many of the new firms were in electronics, computer software and financial services, and several of the largest were US companies looking for sites for European headquarters. Unemployment nevertheless rose sharply, but even during the depths of the recession the local unemployment rate of around 11 per cent remained significantly below the national average of around 13 per cent.

How important was the council's role in attracting these firms? Although many of the incoming firms praised the positive and helpful role of the local authority, it is not clear whether the council was by now playing anything more than an enabling role. Swindon's location, its excellent rail and road links to London and Heathrow, the availability of industrial and office sites, a pool of skilled and unskilled labour, and a good quality environment on the edge of the Marlborough downs, proved powerful attractive forces to firms decentralizing from London and US firms looking for a base for European operations. Swindon's marketing policies were working with a powerful tide, although the town still had to compete with other centres such as Milton Keynes and Peterborough.

Although the main thrust of development policy continued to be aimed at the attraction of outside investment, the rise in unemployment promoted the launching of a number of supplementary initiatives directed more at indigenous enterprise. In 1982 the Swindon Enterprise Trust, a joint council, business and community sector venture, was launched to encourage the establishment of small firms. This was followed in 1983 by the setting up of a local employment sub-committee, providing support for an ITEC and local cooperatives.

The landscape of the town was radically transformed during this period. Office development and the completion of the Brunel shopping centre transformed the central area. Much of the new housing was provided in a series of 'urban villages' on the western edge of the town. The first urban village was built on land purchased by the local authority, but as expansion spread into areas where private builders had substantial landholdings a series of partnership deals became inevitable. Although the Silver Report had favoured a mix of private and public housing (for rent and sale) most of this housing was privately built for owner-occupation. Finally, the peripheral campus estates added concentrations of increasingly futuristic and 'high tech' buildings to the landscape.

Expansion was achieved in spite of delaying tactics, pessimistic forecasts, and sometimes outright opposition from Wiltshire County Council. In 1977 Thamesdown successfully appealed to the Minister about Wiltshire's failure to grant planning permission for western expansion. Thamesdown was also highly critical of the draft North-East Wiltshire Structure Plan when it finally appeared in 1980, Conservative councillors and MPs joining with Labour in opposing constraints. The Secretary of State subsequently approved upward revisions for population, housing and industry, which confirmed government support for Swindon's continuing rapid expansion.

Swindon's apparent 'success' began to attract increasing media interest by the early 1980s. It was during this period that Swindon was dubbed 'the fastest growing micro-economy in Europe', 'riding a high tech wave' (*Financial Times*, 30 June 1983). Growth was not entirely without conflict, however. There were rumblings of discontent from Labour members about some of Smith's more flamboyant pronouncements on the importance of more expensive housing and up-market entertainment in order to attract the captains of industry, and periodic bouts of concern about the extent of the western expansion and the pre-empting of resources needed elsewhere. But, on the whole, the all-party consensus held firm. Significant doubts did not emerge until the mid-1980s.

3 Political change and policy reappraisal since the mid-1980s: the end of consensus?

The mid-1980s seem to have marked a turning point in Swindon's post-war history. Although rapid growth has continued up until the present, this latest period has been marked by greater political conflict, a series of major policy reappraisals, and the questioning of the old consensus behind growth. The causes of these changes are rooted in an array of external and internal pressures, both new and the cumulative product of previous rounds of growth. As we shall see, development pressures have intensified while the resources available to the local authority to control growth have diminished. As the social and economic structure has become more complex, the old certainty that growth was good for all has been eroded. Politically, the period has been marked by more disagreements between the parties, and some conflict within the Labour group reflecting the emergence of Swindon's own version of a 'new urban left'.

The new uncertainties were clearly articulated in the council's consultative document, *A New Vision for Thamesdown*, published in 1984. This document represented the first major reappraisal of growth strategy since the 'Silver Report' in 1968. It seems to have originated from a group of officers who saw it as a means of forcing the politicians to face up to the fact that, because of new economic and social forces, the context of growth in the 1980s was now radically different. Its publication marked the beginnings of a public debate on Swindon's whole future as an expanding town and the role of the local authority in steering future growth.

Up until the late 1970s, the report argued, a 'clear vision' had existed of what Swindon was trying to achieve, there was strong local support, the council was a major landowner, most publicly provided services were under local authority control, there was adequate financial support for development, rate levels were competitive, and private sector development (apart from housebuilding) was just taking off. The council was thus in a position to exert significant control of the overall development process.

All these factors, it was argued, were now changing. The national economic context was more unfavourable, government controls over the local authority's expenditure had tightened, the county council would find it more difficult to fund its share of services for any further expansion, and the provision of basic infrastructure by the public utilities could not be taken for granted. There were also indications that public opinion in some quarters was becoming increasingly hostile to further extensions of the urban area, associating it with increased congestion or threats to the green belt and property values. Given the rising cost of land the local authority would not be able to replenish its land banks and the future provision of jobs and housing would increasingly depend upon the private sector. As a result of these pressures the original vision had become 'clouded' and the local authority was in danger of losing control of the development process.

The report laid out a number of low and high growth scenarios and summarized the means of intervention that still remained for the authority. Following widespread public consultation the Labour council chose a strategy of 'selective intervention', which basically involved an attempt to continue to influence the direction of growth through planning controls, more partnership arrangements with the private sector, and a strengthening of the council's lobbying and advocacy role in relation to central government.

The problem of influencing the future pattern of growth with fewer land and financial resources rapidly became even more acute than the 'New Vision' debate had perhaps anticipated. The local authority was ratecapped in 1985–86 and has remained on the government's list ever since. While the constraints on finance were tightened, the pressures from private developers increased. The problems facing the local authority can be well illustrated in the policy areas of economic development and housing expansion.

Rethinking economic development strategy

Economically, the area benefited strongly from the partial recovery of the national economy from 1983–84 onwards. Although detailed statistics are not available after 1984, it seems that much of the local employment growth was in the service sector, with established companies in insurance and financial services consolidating and expanding their headquarters' functions in the town, and a wide range of new computer service and software firms moving into city centre and peripheral campus sites. At the same time there were important changes in the manufacturing base. Although the British Rail Workshops finally closed in 1986, new high tech firms continued to arrive, and the car industry was given a boost by the construction of an inspection and car engine plant for Honda (with the possibility of a full-scale car plant to follow). Nevertheless, the experience of higher unemployment during the recession of the early 1980s, coupled with the emergence of a stronger left wing grouping within the local Labour party and the election of a more left wing deputy leader, prompted a fundamental reconsideration of economic development strategy.

For some time, there had been growing criticism within the Labour Party of Smith's style and approach to economic development, and following a management review a new Directorate of Economic and Social Development was established in 1986, incorporating an Employment Development Unit. An edition of the Council's own newspaper, *Thamesdown News* (October 1986) explained that economic development policy would be broadened beyond the attraction of new firms towards building a partnership with local community groups, the business community and the labour movement.

The appointment of a new director, Jamie Robertson, immediately stirred up controversy. He had formerly been an economic adviser to the Merseyside Enterprise Board, and had also been a Labour parliamentary candidate. The Conservatives strongly opposed his appointment, which they regarded as an overtly political one. Robertson was asked to produce a new economic development strategy for the 1990s and his subsequent report incorporated a number of 'new left' themes and proposals associated with some of the larger, more left wing authorities such as Sheffield and the GLC. His report pointed out that although the council's promotional efforts had achieved considerable success, incoming employment had been mainly in the service and high technology sectors, and had not been able to absorb the local redundancies in manufacturing because of a mismatch of skills. Far more attention needed to be paid to both the quality of jobs and equality of opportunity. Suggestions included a code of employment practice for organizations receiving financial assistance from the council; contract compliance procedures for suppliers of goods and services to the council; and positive support for employment generation initiatives directed towards women, the disabled and ethnic minorities. It was also proposed to investigate the possibilities of setting up an Enterprise Board and to give particular encouragement to the formation of local cooperatives. Finally, the council should also seek to extend the principle of joint planning by forging closer links with the trades unions and community groups.

These more radical proposals were, however, combined in the report with more traditional ideas. Robertson supported the continuation of Swindon's highly successful promotional policies, albeit in a modified form, and also accepted the need to develop closer links with the private sector through partnership arrangements. Nevertheless, the publication of the report was quickly followed by Smith's resignation and bitter political exchanges between the parties. The Conservative leader warned of 'the thin veil of moderate socialism' in Swindon being 'blown aside' (*Evening Advertiser*, 14 April 1987).

In the event the new strategy did not produce as many changes as were originally envisaged. Officers and councillors admit that in the competitive atmosphere of a third-term Thatcher government it is difficult to impose codes of employment practice on new firms without driving them into the arms of competing towns in the South-East. Although the cooperative sector has expanded, total employment is small, and a proposed Enterprise Board looks as though it will be mainly concerned with filling gaps in the venture capital market. The decline of trades union organization in the town has blocked any moves towards joint planning on any significant scale. Although the promotional element aimed at attracting outside investment has been recast and expenditure reduced, it still remains of central importance. Current advertising still unashamedly carries the slogan, 'Swindon: The Profit Base'.

The pressures of private housing development: the northern expansion

Swindon is highly accessible to those who work in some of the booming local economies of the South-East. The severe restraints on residential development in areas such as Central Berkshire, leading to rocketing house prices in the 1980s, forced more and more households to search further afield for affordable housing. Swindon was the first town, going west along the M4, where housing land was readily available, and the town rapidly became a Mecca for housebuilders. As a result Swindon has become increasingly integrated into the South-East housing market, leading to an inflation of housing prices.

In the mid-1970s housing was cheap, well below the South-West region's average price level. By the late 1980s house prices were approaching the much higher levels of the South-East. This meant that the prospects for manual workers and less skilled white-collar workers, who remained dependent on local wage-levels for the easy access to home ownership that their parents had enjoyed, were reduced. Moreover, the council was able to do little to force private developers to increase the moderate and lower income housing supply. Yet it still had to provide infrastructure and services for the incoming population. Given that it was rate-capped and unlikely to profit much from new development, it was not surprising that there was a decline in enthusiasm for growth and increased questioning of its benefits. However, as the council discovered in relation to the northern development, it could not restrain the growth that it had worked so hard for so many years to promote.

In the 1970s, as we have noted, development took place to the west of the town. It was generally accepted that, once this land had been used up, further growth would be located to the north of the town. Throughout the 1970s astute developers bought up this area, even though it was not shown in the structure plan as development land. In October 1986, a housebuilder's consortium sought planning permission to develop 1500 acres of this land. This appears to have been the largest ever single private application for residential development. Although the plan contained some employment land the basic argument put forward by the developers was that the 9000 houses that they proposed to build, accommodating 24 000 people and expanding the town by 20 per cent, would relieve housing demand in the South-East. As the developer's planning consultant told us, Swindon should now be seen as a 'safety valve for the South-East'. However, the developers were not just planning to build houses, but also to provide parks, a shopping centre and so on. With this new proposal, planning in Swindon moved decisively from being public to private sector led – just the change that the government has sought to promote in the inner cities and elsewhere.

However, this proposal was not welcomed by the district and county councils – who would have to provide the infrastructure and services that would be required. For years the county council had opposed Swindon's growth and the demands it made on the county budgets. But early in 1987 both councils looked set to oppose the development at a public inquiry, arguing that it would divert resources that were urgently needed to improve conditions for the existing population. It is interesting that even the developer's planning consultant told us that, if the government was expecting private enterprise to do the job, additional public resources were needed and it could not just go on 'screwing local authorities into the ground'.

The public inquiry was held in late 1987. Before it took place pressure built up on the councils to do a deal with the developer and withdraw their opposition to the proposal. Few people believed that the developer would fail to get planning approval. Indeed, as the Secretary of State for the Environment was caught between the developers' strident campaign for more housing land in the south and the outrage in the Conservative's electoral heartland at this prospect, the Swindon development offered a political as well as a physical safety valve. In addition, the developers had offered to contribute towards public service provision. But it was made clear that if the authorities did oppose the plan at the inquiry all such deals would be off and they would lose what leverage they had to extract such contributions from the consortium. After various discussions and manoeuvres, in the course of which the district council appears to have used a threat to develop a retail complex on an area of its own land near the proposed development as a lever to extract a better deal, both councils withdrew their objections to the plan in exchange for contributions to infrastructural and other costs. In autumn 1988 planning permission was granted.

In summary, Swindon is now embarked, somewhat more reluctantly, on another round of major expansion, but a round of expansion that is more private sector and housing led than in the past.

Growth for whom?

As a result of these changes the balance between the costs and benefits of growth have become more problematic than in the past. This is not to say that the majority of the population think that life in the town has got worse. Almost 50 per cent of migrants to the town questioned in a recent survey found it a better place to live than they had anticipated; only 10 per cent found it worse.[1] Almost 40 per cent felt expansion had contributed greatly to the range of available jobs. Almost 60 per cent of full-time employees felt that their household finances had improved recently and almost 50 per cent found it 'very' or 'quite' easy to make ends meet, and very few reported that their finances had seriously deteriorated. These attitudes were echoed by households we interviewed informally. Particular praise was lavished on the range of leisure and shopping facilities that the council had promoted. Swindon was seen as a good place to bring up children, 'a new town without the new town blues'.

However, not all have benefited from growth. There has been a mismatch between job losses in the skilled engineering trades and the less skilled jobs that have taken their place. As elsewhere, it has been older workers who have mainly suffered from this change. Also, employment in some sectors is less stable, firms come and go more rapidly than in the past, old skills have to be abandoned and new flexibility in working practices has to be accepted. The extent to which the loss of the society and culture of the railway age outweighs the new opportunities for privatized and family-centred consumption can be exaggerated, but for those left on the sidelines by growth and change the costs often seem considerable and the benefits few. Thus the wife of a redundant rail-worker complained: 'once everybody knew everybody. Now nobody knows nobody . . ., the new people coming in don't seem to think of it as a community. They think of theirselves nowadays. As long as I'm all right, that's what matters.' The pressures of consumerism and debt were also mentioned. 'This is a successful area, so you have got to be successful.' As an unemployed young man, sleeping on friends' floors and living on £23 per week benefit saw it: 'They seem to be recruiting a lot of people from outside, especially with skills, and bringing them all in: and the unemployed who have actually lived here, virtually all their lives, they are finding it difficult to get work because it's all specialized industry.' This man was aware that his dismal job-prospects were matched by the declining likelihood that he would ever be able to afford private housing or gain entry to the town's declining public housing stock.

4 The Swindon experience: lessons from the past and dilemmas for the future

The role of local policy

What conclusions can we draw from Swindon's history concerning the role of the local authority in the transformation of the area? Has the locality been no more than the passive beneficiary, or victim, of wider forces? Certainly the changing balance of 'external' forces has defined the boundaries within which the council leadership has pursued its growth strategy. Up to the 1970s national government played a fairly positive role via the policies of dispersal of industry and population from London and the development of rapid road and rail links. However, this does not mean that Swindon's growth was simply an inevitable outcome of government policies and market forces. Up to the 1970s, at least, the local commitment to growth certainly did make a considerable difference.

Here, it is interesting to compare Swindon with two other towns which considered using the Town Development Act to restructure their local economies – Ashford and Banbury (Harloe, 1975; Stacey, 1960; Stacey *et al.*, 1975; Brown *et al.*, 1972). In Ashford and Banbury, in contrast to Swindon, the expansion plans led to conflict. The differing politics and social structure of the three towns contributed to this outcome. Labour controlled Swindon almost permanently, but anyway there was virtual unanimity on the necessity for expansion and diversification of the local economy and a willingness to accept London overspill as a means of achieving this aim. The two other towns had a larger, more politically important middle class, which saw little of benefit to itself in an influx of industry and population from London. Town development was therefore resisted and the opportunity to use it for local regeneration passed by.

However, Swindon's success required more than the political resource of a growth coalition supported by all the significant local socio-economic and political elites. It also required human resources and organization. The growth policy enabled the town to attract some able officials, who pursued the goal of development with competence and singlemindedness. Development became not just another local authority function, it lay at the heart of almost everything that the council did and dominated its organization, decision-making and politics. In the early years of development, when Swindon had to badger and even sometimes outwit those who had access to the resources it needed for growth, the political skills of this small group of officials were of central importance. Later – when Swindon moved from being 'an expanding town' to 'a new city' – the skills required became more technically complex and varied, but now the very success of development meant that the council could employ a large and dynamic staff, almost a new town development corporation within the structure of a local authority.

As we have seen, however, now that the local authority's powers and resources have become more circumscribed and the market has decisively taken over from the public sector as the main motive force in growth in the 1980s, the local political elite is losing the resources that enabled it to bend the forces of development to serve its conception of local needs. There is now a more questioning attitude to further development, and a wish by the council to give higher priority to social development and improvements in the older parts of the town. But with declining resources available from development and from central government it has become increasingly difficult even to sustain existing levels of service and public employment. Now, as the opportunities for 'creative accountancy' are running out, pressure on the council to consider asset sales to finance future expenditure has intensified. But, in the long run, this will have the consequence of reducing the revenue from earlier development, so restricting what the local authority can do even further and forcing it to rely, as the 'New Vision' report suggested, on its more conventional powers to attempt to influence the way in which further growth occurs. As a result of these changes the political and organizational basis of the growth coalition is under increasing threat.

Models of local politics

The Swindon experience is also relevant to several broader issues concerning the nature of local political systems. We have argued that the construction of a local growth coalition was important in Swindon's expansion. Much has been written on such coalitions, mainly in an American context (but see Pickvance, 1985; Bassett, 1986). However, what we might describe as the active components of the Swindon coalition were differently constituted from those that appear to have been typical in American cities (see, for example, Logan and Molotch, 1987). Local business interests have not been nearly as actively involved in the politics of development in Swindon as in many of its US counterparts. Even today the Chamber of Commerce has only a marginal influence, and tends to represent only the interests of small-scale local capital. Friedland (1982) has also suggested that organized labour has strongly influenced policy in some US cities. Trades union influence in Swindon, however, has been more indirect and tended to decline over time. Many of the railwaymen on the Council after the war were also trades unionists, but the economic restructuring of the past few decades has resulted in a serious decline in trade unionism, outside the public sector. The representation of the local union leadership on the council, at one time very significant, has now almost disappeared, and the local trades council appears to exert little influence even on a Labour council. The only strong area of increased trade union activity has been in the public sector, and here the main priorities have tended to be the protection of jobs and existing services rather than growth related objectives. But, crucially, both business and unions have endorsed the objectives of the relatively small group of individuals who have coordinated the growth policy.

So although the concept of a 'growth (or spatial) coalition' provides at least a useful description of, if hardly a 'theory' about, what has occurred in Swindon, the range of its active components should not be exaggerated. In reality there was a strong coincidence between the political interests of key local politicians, especially in the Labour party, and the professional interests of officials in supporting expansion. For many years this group was able to sustain a form of 'hegemonic project', the fundamental basis for this being the promise that local authority guided expansion would provide increased jobs, housing, retail and leisure facilities, which would benefit almost all sections of the community.[2] It is this material basis to political control that is now under threat.

The Swindon experience is also of some relevance to recent debates about the possibilities for 'local socialism'. The local Labour party has, by and large, been a traditional 'moderate' party, which has adapted itself to the rather unusual situation, for Labour, of being in control of a fast-growing, increasingly white collar dominated, medium-sized town in the heartlands of southern England. The strategy pursued for much of the post-war period can be described as one of using public enterprise to support and guide market forces towards a preferred pattern of economic growth and, through the expansion of public services, to distribute the benefits of that growth more widely and more equitably than would otherwise have been the case. Although councillors have regarded this strategy as essentially socialist it would perhaps be more accurate to describe it as a form of 'municipal labourism', stopping well short of any attempt to transform the social relations of production in the local economy or restructure the local state machine in the interests of greater democracy. Although there was a period of turmoil in the mid-1980s, when Swindon's own version of a 'new urban left' seemed to be emerging, the central elements of the strategy survived. It survived because, in its own terms, it was remarkably successful. It was successful because of Swindon's drive and its location, which enabled it to take advantage of the broader processes of restructuring in the South-East region.

However, this attempt to combine economic growth with progressive social policies may no longer be a possible strategy for Labour to adopt. Although the pressures for growth continue to increase, the powers of the local authority to intervene to control and shape development have been decreased. The old consensus has been shaken and local Conservatives are convinced that the pattern of economic and social change is moving in their favour. Labour's challenge is to try to construct a new progressive coalition in an increasingly middle-class community. A new, pragmatic, 'soft-left' leadership has recently emerged, which recognizes these new problems and places its faith in asset disposal to raise finance, persuading local capital to invest more in social facilities, and widening its support by a policy of 'open government'. The chances of success are not high, and the dangers are that a new growth coalition may come increasingly to resemble those that exist in the United States, an alliance of some local politicians, the middle class and business, pursuing their own sectional interests with little concern to link economic growth to redistributive social policies.

Acknowledgements

We gratefully acknowledge the help of our co-researchers on the Swindon project – Martin Boddy, Gill Court, John Lovering and Jane Wills – and all those who helped us in Swindon.

Notes: Chapter 2

1 This survey was carried out for a parallel ESRC research initiative, Social Change and Economic Life.
2 Jessop (1983) has developed this concept to refer to the mobilisation of support behind action which asserts a general interest in the pursuit of objectives that implicitly or explicitly advance the interests of a leadership group. Recently Cox and Mair (1988) have applied this concept to the analysis of business dominated growth coalitions in the USA. All these authors are referring to capitalist led hegemonic projects but the concept may also be useful in analysing the rather different circumstances discussed in this chapter. As this book was going to press Hajer (1989) has published an account of urban politics in Oxford, which also seeks to apply the concept at a local level.

References: Chapter 2

Bassett, K. 1986. 'Economic restructuring, spatial coalitions and local economic development strategies, a case study of Bristol', *Political Geography Quarterly*, Supplement to **5** (4), 163–78.
Boddy, M. and Fudge, C. (eds) 1984. *Local socialism?* London: Macmillan.
Brown, T., Vile, M. and Whitemore, M. 1972. 'Community studies and decision taking'. *British Journal of Political Science*, **1** (2), 133–53.
Cox, K. and Mair, A. 1988. 'Locality and community in the politics of local economic development'. *Annals of the Association of American Geographers*, **78** (2), 307–25.
Friedland, R. 1982. *Power and crisis in the city*. London and Basingstoke: Macmillan.
Gyford, J. 1985. *The politics of local socialism*. London: Allen and Unwin.
Hajer, M. 1989. *City politics: hegemonic projects and discourse*. Aldershot: Avebury.
Harloe, M. 1975. *Swindon: a town in transition*. London: Heinemann Educational.
Jessop, R. 1983. 'Accumulation strategies, state forms and hegemonic projects', *Kapitalistate*, **10–11**, 89–111.
Levin, P. 1976. *Government and the planning process*. London: Allen and Unwin.
Logan, J. and Molotch, H. 1987. *Urban fortunes. The political economy of place*. Berkeley, Calif.: University of California Press.
Ministry of Housing and Local Government 1964. *The south-east study 1961–1981*. London: HMSO.

Pickvance, C. G. 1985. 'Spatial policy as territorial politics: the role of spatial coalitions in the articulation of "spatial" interests and in the demand for spatial policy'. In G. Rees *et al.* (eds), *Political action and social identity: class, locality and ideology*. London: Macmillan.

Stacey, M. 1960. *Tradition and change. A study of Banbury*. London: Oxford University Press.

Stacey, M., Batstone, E., Bell, C. and Murcott, A. 1975. *Power, persistence and change. A second study of Banbury*. London and Boston: Routledge and Kegan Paul.

Swindon Expansion Project. Joint Steering Committee 1968. *Swindon. A study for further expansion*. Swindon: The Committee.

Thamesdown District Council 1984. *A new vision for Thamesdown*. Swindon: The Council.

3

Trying to revive an infant Hercules: the rise and fall of local authority modernization policies on Teesside

RAY HUDSON

1 Introduction

In 1987 Mrs Thatcher appeared in national newspapers and on the nation's TV screens, posing against the background of some of the devastated old industrial areas on Teesside. Her visit there drew attention to two things. First, the changed political climate that underlaid new central government policy responses, such as Urban Development Corporations, to the problems of de-industrialization in inner areas of conurbations. As she said at the time: 'We are setting out again to be ahead of our time . . . where you have initiative, talent and ability, the money follows'. Secondly, it provided a graphic illustration of the dereliction, unemployment and poverty that have swept through the area in the last decade as a consequence of industrial decline. The pictures evoked memories of the 1930s on Teesside in places such as Stockton. Depression in the 1930s and 1980s contrasted powerfully with the major expansive boom of the nineteenth century and with the hope and optimism, in and for Teesside, in the early 1960s. Not so long ago it seemed that Teesside could anticipate a prosperous future, one of full employment and rising affluence. Centre-stage in this image of the future stood central and local government, integrally involved in transforming the area via implementation of a modernization project, for which there was very broadly based support from trades unions and private sector alike, and across the whole party political spectrum. The legacy of the old nineteenth-century economy was to be radically reconstructed. The new economy would still be focused around the chemicals and steel industries, but these would be transformed by massive state-subsidized investment to provide a centre of technologically sophisticated modern manufacturing industry. New industries would diversify employment opportunities. Furthermore, the built environment would be transformed by local authority planning and expanded

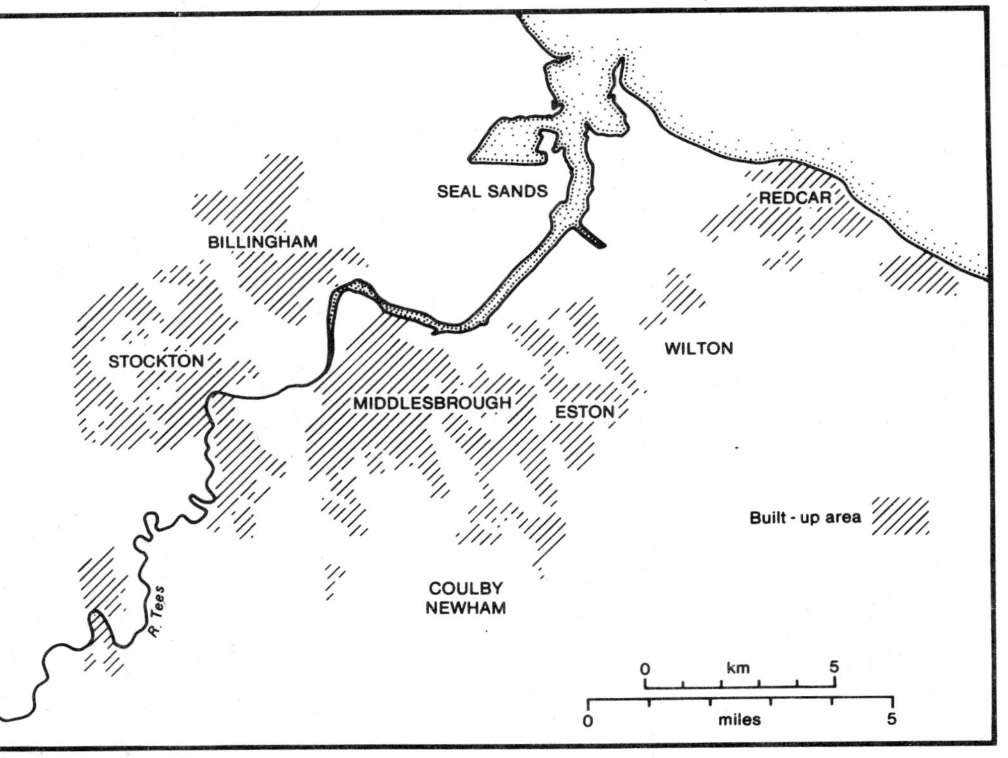

Figure 3.1 Teesside.

public expenditure on roads, houses, shopping centres and educational and health facilities. Modernization was seen as embracing all aspects of economic and social life.

In the early 1960s, Teesside seemed destined for a long boom. It occupied a central place in the strategies of major chemical and steel producers to remain profitable and internationally competitive. It was also a pivotal location in political strategies for regional and national modernization. Modernization of industries such as chemicals and steel, which produced inputs for many manufacturing activities, was vital to the emergent national strategy of restructuring manufacturing in the UK into an internationally competitive sector. Teesside itself was specifically identified as a key location in the Hailsham White Paper (Board of Trade, 1963) proposals to modernize the North-East. Thus, Teesside occupied a unique niche at the intersection of national and regional modernization policies. A powerful consensus, uniting the interests of key sections of capital and labour and their political representatives in Conservative and Labour parties, informed local authority policy-making. It was associated with the 'getting things done' mentality, described by Gladstone (1976, p. 50) in these terms:

> 'Getting things done' depends on an argument that goes as follows. The first priority is bringing jobs to the area; being able to attract employers depends on their attracting bright young executives. And this in turn depends on being seen to do 'prestige' projects which get you into the twenty first century. And this is largely a matter of building large scale city centre development and a system of urban motorways.

Thus local authority policies were shaped to meet the needs of commercial and industrial capital that might be attracted to Teesside, to complement those of central government and the big corporations on whose decisions the implementation of modernization policies ultimately depended. The alliance between these companies, the trades unions whose members found or retained employment with them as fresh capital flowed in, and the local councils, many of whose members were employed by these companies, was extremely powerful. It cut across class boundaries in a vigorous promotion of one conception of what Teesside's future ought to be. Consequently, local authority land use and infrastructure provision policies reflected the requirements of the major chemicals, oil and steel companies and those of commercial and financial capitals seeking lucrative city centre sites, and they usually evolved in this way without obvious pressure being directly exerted by these capitalist interests themselves. Modernization was a truly hegemonic project.

Implementing this modernization project, especially in the absence of a regional tier of government, would crucially depend upon local authorities in the Teesside area. They would need to be centrally involved, using the powers available to them and deploying the various resources at their disposal. Indeed, both the organization of local government and the content and forms of local authority planning were altered to allow more effective action, initially as part of a local response to a perceived need for change, later as part of a national reorganization of local government and planning. A complicated pattern of local authority boundaries cut across the area and so several councils were involved in the process of implementation. Furthermore, some authorities were under Labour control while Conservatives controlled others. The changes that were made partly reflected the problems arising from the divergent competitive interests of local authorities (especially between 'rural' Conservative and 'urban' Labour controlled authorities) and partly those arising from a lack of technical expertise to cope with the demands of the new planning approaches. Despite these changes, however, there were strict limits to the scope for autonomous local authority action, arising from their place in the overall structure of the state, from the priority attached to national rather than local objectives, and from the overall limits to state activity in a capitalist society.

Consequently, implementation of the hegemonic modernization project was problematic. Initially, opposition to it centred around the distribution of costs and benefits within Teesside resulting from its implementation, as some social groups and areas reaped the benefits while others bore the costs. For example, as it became clear that industrial modernization policies produced pollution and a threat to the environment, opposition to them began to crystallize around an emergent 'green' politics. This deepened as it later became clear that the intended policy effects were no longer being achieved, although the unintended and unwanted ones continued. Marked disjunctions between intentions and outcomes began to emerge. It became clear that there were forces beyond local authority control or regulation producing changes in and through Teesside. As local authority policy objectives came into conflict with those of central government, the latter took precedence. To a large extent, this conflict reflected government attempts to grapple with the growing difficulties of managing the trajectory of national economic development in an increasingly internationalizing economy. Not surprisingly, therefore, the experiences of Teesside have painfully revealed the limits to local authority effectiveness in implementing policies in the face of decisions by nationalized industries and major multinationals, driven by the imperatives of a changing global economy and geography of production and trade in bulk chemicals and steel. Under these circumstances, the consensus that had been formed around the hegemonic project of the 1960s, rooted in a statist politics of modernization, came under increasingly severe strain. As the national political climate altered in the 1980s, this was reflected in Teesside in further emphasis, through local authority policies, upon the encouragement of small firms, however unlikely an environment the area provided for the flowering of an enterprise culture. At the same time, central government became increasingly involved in the area by the designation of Enterprise Zones, its joint involvement with the local authorities' Urban Programme, and by the creation in 1987 of an Urban Development Corporation. The cosy consensus that existed in the early 1960s has been replaced by a rather uneasy tension, as considerable uncertainty hangs over Teesside's future and there seem to be no possibilities for constructing a new hegemonic project, within the limits of a new-right political economy, that could adequately respond to the problems confronting the area.

In the rest of this chapter, I shall elaborate upon these introductory remarks. First, some consideration is given to events on Teesside prior to the early 1960s, to Teesside's changing role in the international economy, and the role of the state in these events. This provides a necessary background to the emergence of the hegemonic modernization project in the 1960s. Secondly, the changed forms of planning and local government reorganization on Teesside are examined a little more fully. Next, the formulation and implementation of modernization policies for commercial redevelopment and for manufacturing are described and contrasted. Finally, some conclusions are drawn from Teesside's experiences.

2 A historical sketch of development on Teesside

Teesside was one of the boom areas of nineteenth-century industrial capitalism. It occupied a pivotal location in the accumulation process as key commodities and burgeoning profits were produced in and through it. This expansion was centred on and exemplified by the growth of Middlesbrough, which increased in population from a handful of people in 1831 to more than 100 000 in 1901, with massive in-migration from other parts of the North-East and other regions of the United Kingdom. As Gladstone remarked, on a visit to Middlesbrough in 1862, it was 'a remarkable place, the youngest child of England's enterprize . . . It is an infant, gentlemen, but it is an infant Hercules' (cited in Briggs, 1963, p. 241). This boom town growth began with the creation of port facilities to ship coal from the rapidly expanding Durham coalfield, but was fuelled most strongly by the development of the iron and steel industry. This involved iron ore mining and iron and later steel production. The main capitalist interests involved in these activities, the Teesside Ironmasters, quickly dominated the national market, but at the same time iron and steel production on Teesside was tied into an internationalized economy virtually from the outset. By the last quarter of the nineteenth century there were considerable imports of iron ore into Teesside from northern Spain while much of Teesside's iron and steel output was exported, both directly and indirectly. Shipbuilding and heavy engineering companies developed, which used the outputs of the steel industry as a major part of their own inputs, and were heavily involved in international markets. Many of these activities, connected by physical input–output links also became connected in terms of ownership and control through interlocking share-ownerships, mergers and so on.

Around these booming industries there developed a series of settlements, built with no other purpose than to house the workforces at minimum cost. The construction of these working-class housing areas is graphically described by Lady Bell, the wife of one of the Middlesbrough Ironmasters (1970, pp. 2–3). There can be no doubt that Lady Bell was well placed to comment upon the process, for the Ironmasters and the other major industrial capitalists dominated most aspects of life in the town. This pervasive paternalism left a deep mark on the emergent working-class political culture, for there was little alternative to employment 'at the works'.

At the same time, across the River Tees, developments were taking place in a newer industry that had a profound impact on Teesside's future growth and decline. There had been some development of chemicals production in the latter part of the nineteenth century, but this took on new dimensions as Brunner Mond, from 1918, and Imperial Chemical Industries (ICI), from 1926, developed a vast inorganic chemicals complex (copying those of the major German companies, such as IG Farben, in a way that anticipated BSC's mimicking of the major Japanese steel companies some fifty years later). Another company town began at Billingham to house the workforce of this complex, many of whom moved there from the Durham coalfield in the inter-war years (Pettigrew, 1985, p. 126). Although parts of Teesside were ravaged by the effects of recession on the steel producing and consuming industries in this period (for example, see Priestley, 1934), the growth of ICI's activities meant that there were areas of comparative prosperity and economic dynamism on Teesside.

After 1945, Teesside continued to be a centre of growth in fixed capital investment, output and employment. Demand for its more traditional commodities, steel and heavy engineering, revived and that for fertilizers grew. Furthermore, ICI's Teesside operations were greatly expanded following the development of the Wilton petrochemicals complex (Hudson, 1982). As ICI were granted deemed planning permission by central government for this major 2000-acre site, local authorities had little influence on the pattern of its development. Fears of ICI experiencing labour shortages in what was seen as a key national sector led to a deliberate embargo on alternative new manufacturing employment for men being introduced into the area. The contracting iron ore mines in the eastern part of Teesside were designated as a source of labour supply for Wilton. Industrial estate provision to provide employment for women – diversifying employment – began to expand, however (North, 1975, pp. 111–14). Moreover, the local authorities now took on the main responsibility for providing working-class housing for key workers and more generally.

This concentration of economic growth in Teesside in the 1950s assumed a particular importance as the Hailsham programme for the modernization of the North-East was evolved in the early 1960s (Hudson, 1983). Successful past growth and the prospect of this continuing provided compelling reasons for designating Teesside part of the 'growth zone'. Furthermore, such proposals coincided with the views of the Teesside Industrial Development Board. This originated in the 1930s but flourished after 1945 as an important cross-party forum, incorporating representatives of most local authorities, local chambers of trade and commerce, and major trades unions. It advocated growth through modernization of the Teesside economy. Thus there was a receptive and politically broad-based climate of opinion on Teesside that welcomed the Hailsham proposals. It was recognized, however, that further growth on Teesside could provide serious urban planning problems; an unpublished draft of the Hailsham report clearly acknowledged that the projected extent and form of future growth there was based on untested premisses and hence could not be predicted in a way that made it amenable to reliable forecasting and planning (Carney and Hudson, 1974). Nevertheless, the published Hailsham programme suggested that Teesside required a comprehensive urban planning framework to provide guidelines for implementing modernization policies and a mechanism through which to guide and manage growth. It also suggested what the content of modernization policies on Teesside ought to be: construction of roads to improve access to and links between industrial, commercial and residential areas; construction of new industrial estates; and provision of land for chemicals and steel industries by opening up the Teesmouth area and reclaiming derelict sites.

3 Modernization policies, new forms of urban planning and new institutional arrangements on Teesside[1]

The necessity for new forms of urban planning was seen as sufficiently urgent for the Hailsham White Paper to suggest that central Government partly finance the task of devising these forms. Achieving desired environmental improvements was seen as partly dependent upon technically more sophisticated urban planning, more in tune with the demands of both the time and place. The point was made as follows (Board of Trade, 1963):

> The Minister of Housing and Local Government proposes to discuss with them [local authorities] and other authorities concerned the commissioning, with Exchequer assistance, of a comprehensive Teesside Survey and Plan . . . the survey would . . . be the first comprehensive project to be initiated in the light of the Buchanan and Crowther Reports on the Long Term Problems of Traffic in Towns.

Such a Plan would involve collaboration between the numerous Teesside local authorities that then had planning responsibilities, other public agencies (such as those responsible for rail and road transport) and relevant central government departments. There was an initial reluctance by the local authorities to engage in the joint planning exercise[2] (see the foreword by Sir Charles Allison, chairman of the steering committee, to Wilson *et al.*, 1969). However, the widespread support for proposals that would single out Teesside as a major centre of investment and modernization over-rode the more parochial pressures that initially lead to such reservations. In 1964 a steering committee and a supporting technical committee were created. Committee members were drawn from organizations with interests in the construction of the Plan, to provide a mechanism through which coordination could occur.

Consultants began work on a Teesside Survey and Plan (Wilson *et al.*, 1964). Significantly, the consultants saw themselves as preparing a land-use/transportation plan. They considered seven possible development strategies. Their chosen option owed much to the need to temper optimal planning considerations with the requirement to minimize public expenditure (see Teesside County Borough Council, 1972, pp. 15–17). The Teesside Survey and Plan (hereafter *Teesplan*) was completed in 1968 (Wilson *et al.*, 1969), coinciding with the passage of the new Town and Country Planning Act. The 1968 Act was intended to reorganize the basis of local authority planning, increase the efficiency of the plan-making process and bring it more into line with the requirements of a modern capitalist society. The new system would be two-tier. Local and Action Area Plans would guide development control, integrated within an overall framework provided by a Structure Plan, which would specify a broad longer-term land-use *and* socio-economic development strategy. The reference to socio-economic issues represented the most significant change from the previous Development Plan system. Although *Teesplan* subsequently served as an advisory document in the process of Structure Plan preparation, there were inherent limitations in its capacity to fill this role adequately. It was essentially a land-use/transportation plan, but Structure Plans were now required to be much more than this.[3]

Although *Teesplan* was widely believed to be a draft Structure Plan, there was an immediate complication arising from specific local government reorganization on Teesside, involving the creation of Teesside County Borough. These local changes were then overtaken by more general ones, reorganizing local government into bigger units. The Royal Commission on Local Government argued for an expanded Teesside local government area. This provoked fierce resistance from existing local authorities, who would lose out as a result. After considerable conflict (especially between urban and rural authorities) and much debate, a compromise emerged. The new County of Cleveland was created in 1974, made up of four constituent boroughs: Hartlepool, Langbaurgh, Middlesbrough and Stockton. As well as a Teesside Structure Plan, preparation began on Structure Plans for Hartlepool, East Cleveland and West Cleveland.

Despite the complex changes that resulted from reorganization of local government and changes in party political control of local government on Teesside in this period, from Labour to Conservative to Labour again in 1974, there was a marked continuity in planning and corporate policy objectives. This was reflected in the continuity between *Teesplan* and the subsequent Structure Plans. Both major political parties shared a powerful commitment to modernization policies, while changing party political control allowed senior officers, as technical experts, to exercise considerable influence over the planning process. Put another way, modernization was affirmed as a hegemonic project on Teesside, as in the North-East more generally during this period.

4 The formation, modification and implementation of local authority modernization policies

The main emphasis in *Teesplan*, as in the Hailsham programme and in *Challenge of the Changing North* (Northern Economic Planning

Council, 1966), was upon modernizing and strengthening Teesside's economy.[4] There was also a considerable emphasis in *Teesplan* on local authorities' supporting the 'local economy', for example, providing specific sites and premises, and absorbing part of the costs of reproducing labour-power. For local authorities to successfully undertake this range of activities would require them to dovetail their activities and expenditures with those of other public and private sector agencies. Particular emphasis was placed upon meeting the requirements of 'capital intensive' chemicals and steel producers. This prioritization reflected the powerful representation of chemical and steel interests on local councils, through the election of their employees as (usually Labour party) councillors,[5] and a recognition of the strategic national importance of these industries on Teesside. However, it was widely recognized that major expansion of capacity and output in these industries would be accompanied by, at best, gently falling employment (Beynon *et al.*, 1985; Hudson, 1982). The *quid pro quo* for this, however, was a political commitment to provide alternative employment in manufacturing and services. This was reflected in the promise of local authority resources to help diversify employment opportunities.[6] As long as employment in chemicals and steel remained buoyant, however, pressures for alternative manufacturing jobs remained muted. By the late 1960s, for example, there were only 12 000 jobs on all industrial estates on Teesside (North, 1975, p. 111).

By the time *Teesplan* had been succeeded by the Teesside Structure Plan (Cleveland County Council, 1977), the overall economic climate had altered sharply and unemployment was rising. However, the sequence of documents involved in preparing this Structure Plan merely emphasized the degree to which analyses of the causes of Teesside's economic problems and resultant policy prescriptions remained rooted in a view of the 1960s that bore less and less relevance to the realities of the 1970s.[7] This was nowhere more clearly exemplified than in the way in which the *Report of Survey* (Teesside County Borough Council, 1972, para. 6.4) repeated verbatim the analysis of the problems of the Teesside economy given in the 1963 Hailsham White Paper.

Industrial land provision

In 1970, chemicals and steel accounted for about half of the industrial land on Teesside, some 6000 acres, equivalent to about one-eighth of the entire area (Teesside County Borough Council, 1973, p. 29). Even so, there were growing demands for more land to be allocated to these uses. Both BSC and ICI had extensive holdings of undeveloped land to cater for their own future planned expansions (Beynon *et al.*, 1986;

Hudson and Sadler, 1985). There was, however, a marked shortage of suitable sites for other chemical companies that might want to locate on Teesside. This was clearly recognized in the Teesside Structure Plan *Report of Survey* (Teesside County Borough Council, 1972, p. 43), which argued that the only large area available for such developments on Teesside was at Seal Sands. Pressures to reclaim land at Seal Sands had already been intensified by the Regional Ports Survey (Northern Economic Planning Council, 1969). This recommended that the Tees should be the focus of major port development in the Northern Region and become a prototype for a new central government plan, the Maritime Industrial Development Area Scheme (MIDAS), linking major new port facilities with complexes of bulk processing and associated industries (National Ports Council, 1972). The MIDAS proposals were abandoned, but they helped shape views about the most appropriate development strategy for Teesside, which continued along the lines mapped out by MIDAS, for reasons that were more related to national than to local needs (Cleveland County Council Planning Department, 1975, p. 59).

Reclamation of 1000 acres of the Seal Sands site proceeded quickly. Responsibility for this lay with the Tees and Hartlepool Port Authority (THPA), although local authorities began to take a growing role in programming reclamation and land management. Several American-based oil and chemicals multinationals began to establish capacity there.[8] The major impetus to the development of Seal Sands was given, however, by the discovery of North Sea oil. A considerable consensus soon emerged on Teesside, centred on the perceived desirability of encouraging North Sea oil and related developments. The THPA strongly favoured this because of its implications for the future of the port and proposed reclaiming land at Seal Sands specifically for this purpose. Billingham Urban District Council also favoured this because of the consequent boost to its rate income. A Working Group was set up in 1971, made up of representatives of the main groups involved or interested in the future of Seal Sands, including conservationists, industrialists and the local authorities (Teesside County Borough Council, 1972). While conservationists raised questions about environmental impacts, the local authorities expressed optimism about the benefits that oil-related developments would bring to the local economy but recognized that such developments would involve 'relatively few jobs' (Working Group on the Implications of North Sea Oil, 1972, p. 43), as the early 1970s euphoria that led to projections of considerable numbers of new jobs evaporated (Cleveland County Council Planning Department, 1979a). Despite the reservations that were expressed, the proposals were accepted and incorporated in the Teesside Structure Plan. Even so, Cleveland County Council soon proposed allocating more land for 'labour intensive' manufacturing activities and for 'small firms'. As the County Planning Department (1975, p. 64) argued, in view of the dangers and problems inherent in promoting chemicals, oil and steel expansion, 'it might be argued that diversification of employment should be the *main* objective of structure plan industrial policy' (emphasis added). These proposals to give greater priority to other forms of industrial development were strongly and successfully resisted by the powerful lobby of the THPA, the major private sector companies and trades unions. This seriously compromised implementation of local authority policies to diversify employment opportunities.

The Seal Sands development was the outcome of a partnership between various agencies of the UK state, the European Community and private capital. The main element in it was the Phillips Norsea oil terminal and separation plant (Hudson, 1982). The THPA initiated much of the reclamation work and the county council took responsibility for coordinating activities via its development control and project management functions. The council also partly funded the land reclamation and provision of services, but much of the finance for drainage, power supply and roads came from the European Community's European Regional Development Fund. The water authorities, through their controversial investment in the ecologically sensitive Cow Green reservoir (Gregory, 1976), and later that at Kielder (the latter again substantially financed by the ERDF), ensured more than adequate water supplies (Beynon *et al.*, 1989). Finally, regional policy assistance from the UK state substantially underwrote the fixed capital investment costs of multinationals located there (Hudson, 1982). From one point of view reclaiming and developing Seal Sands was an impressive example of successfully implementing state-coordinated modernization policies. Even so, this was not achieved without controversy. Growing disquiet about the project was instrumental in generating a more general critical appraisal locally of the sorts of local authority policies that it typified: first, because growing opposition from conservation and environmental groups reflected the emergence of a strengthened 'green' politics on Teesside; secondly, because of the minimal job-creating effects of the development despite its massive subsidization.

During the mid-1960s growing concern about pollution arising from chemicals and steel production on Teesside was effectively suppressed by local politicians, many of whom had close connections with the major companies involved (Gladstone, 1976, p. 44). By the mid-1970s, growing environmentalist opposition to further reclamation was not so easily suppressed. The conflict about the environmental effects of catering for port-related industrial development continued to simmer, occasionally flaring up and becoming more complicated. For example, in modifications to the Hartlepool Structure Plan in 1978 (Cleveland County Council, 1979), the Secretary of State for the Environment proposed making an Area of Special Scientific Interest, adjacent to the Seal Sands area, available for port-related industries. The main trades unions and employers' organizations closed ranks on a 'jobs not birds' ticket.[9] In contrast, both Hartlepool Borough Council and Cleveland County Council objected to these proposals on environmental grounds and because no case had been made for increasing the acreage already zoned in the Teesside Structure Plan for port-related industry.

So in the 1970s concern steadily grew among the local authorities, especially among some of their officers, about environmental pollution and safety hazards (Cleveland County Council Planning Department, 1979b), associated with land-use policies that prioritized the needs of major chemical, oil and steel developments. The consensus on modernization policies began to be questioned, not simply by the 'green' politics of environmentalist groups but also from within the local authorities, especially as jobs were increasingly lost from these industries. The 'jobs at all costs' philosophy, which had been used to justify policies supporting such industries, became increasingly difficult to sustain. This questioning of the appropriateness of a key strand of the 1960s' hegemonic project led to calls for a re-evaluation of local authority land-use planning and economic development policies. Policies that encouraged polluting industries were increasingly seen as conflicting with those that sought to diversify industrial employment.

Pressures for an enhanced programme of industrial estate and advance factory provision to diversify employment opportunities grew from the mid-1970s along with unemployment. Cleveland County Council (1977) was acutely aware of the problem of unemployment and of its causes in the character of chemicals and oil developments and the restructuring of BSC's operations on Teesside. It continued to reiterate its commitment to diversification, although employment growth on industrial estates in Cleveland had been very modest – 7400 net in the period 1972–7, with only 11 400 new jobs created by companies moving into Cleveland between 1965 and 1976 (Robinson and Storey, 1981). To help put these figures in perspective, there was a gross decline of 34 000 in employment in Cleveland during this 11-year period. Nevertheless, the Structure Plan *Report of Survey* had forecast that an additional 31 100 jobs in 'labour intensive' manufacturing would need to be provided in the period 1969–91,[10] for which 1360 acres of land 'must be made available' (Teesside County Borough Council, 1972, p. 117). The County and the four Borough

Councils therefore sought to increase industrial estate provision, much of it funded by the European Community (Cleveland County Council, 1980). However, in practice the problem has not so much been a shortage of industrial land (as structure plan forecasts had suggested) but rather a shortage of private companies responding to the availability of such land on industrial estates: 'the County appears to have more than enough land to meet demand to the end of the century and beyond' (Cleveland County Council, 1985a, p. 4).

Recognizing the limits to a policy of industrial land provision alone, from the mid 1970s the borough councils began building small advance factory units, as did BSC (Industry) in Hartlepool, to complement the English Industrial Estates Corporation's emphasis upon larger units. The boroughs have also encouraged private sector developers to build small units for sub-letting to smaller firms. Specific structure plan policies to encourage small local firms only appeared in the Hartlepool Structure Plan, but their presence there indicated a broader shift in policy (Cleveland County Planning Department, 1979a, p. 4), which was partly a reaction to a critique of the externally controlled branch plant economy (Firn, 1975).

Following lobbying in 1976–7, Middlesbrough had been accorded Programme Authority Status under the 1978 Inner Urban Areas Act, which allowed it to provide some very restricted financial assistance to small firms. From 1979, Cleveland County Council also became increasingly involved in seeking to devise alternative policies to provide direct financial assistance to small firms (Gallant, 1982). Because these policies were funded via section 137 of the 1972 Local Government Act, they also were limited. By 1984 the county council had spent more than £3.0 million on almost 30 different schemes, allegedly saving 900 existing jobs and creating 2200 new ones (Cleveland County Council, 1985a).

Like their counterparts elsewhere (Hudson and Plum, 1986), and in competition with them, local authorities in Cleveland were devising 'autonomist' industrial development policies, ranging from publicity and promotion to financial aid to private companies. Within the tight limits prescribed for it (but which it also endorsed), the county council was imaginative and willing to try out new policy initiatives. However, these had only a marginal impact in stemming the rising tide of unemployment. Small firm policies only generated 3700 jobs between 1965 and 1977 (Robinson and Storey, 1981). Indeed, the county council Planning Department (1979a, p. 28) recognized this at a comparatively early stage. As in the case of the further provision of land for chemicals and oil development, there was a section of opinion within the local authority that questioned the efficiency of existing and emerging policy initiatives. Yet the county and borough councils continued vigorously to pursue small firm policies – for example, by successfully lobbying for the designation of an Enterprize Zone in Middlesbrough to add to that which had existed since 1981 in Hartlepool. Local politicians felt the need to be seen to be doing *something* about unemployment and could see no alternative to small firm policies (a view shared by some academic commentators: for example, see Robinson and Storey, 1981). Although such policies had little effect in creating employment, they did have a considerable ideological effect in legitimating Thatcherite solutions to local economic problems. Promoting such policies lent credibility to the suggestion that unemployment was to be removed by means of local authorities competing among themselves to help create an enterprise culture.

Commercial redevelopment and the promotion of central Middlesbrough as a sub-regional centre

In the early 1960s there were local authority proposals for town centre redevelopment in Middlesbrough. These were rejected by the Ministry of Housing and Local Government, which argued that a more comprehensive town centre plan was needed. Such proposals soon emerged in the Hailsham White Paper. This placed great emphasis on town centre redevelopment, partly reflecting pressures from commercial and finance capital for new investment opportunities to be created outside the 1950s property boom area of London (Marriott, 1967), from the 'road lobby' (Hamer, 1974) and other construction interests, and from professional planning organizations. The White Paper argued that town centre redevelopment would help to diversify employment opportunities, both directly and indirectly, by presenting a more favourable image to potential in-migrant manufacturing companies. Although particularly emphasizing the redevelopment of central Newcastle in its role as regional capital (Regional Policy Research Unit, 1979, Part 9), it also advocated similar changes in central Middlesbrough so that it would become a commercial and social focus for the Teesside conurbation. Following publication of the White Paper, Middlesbrough County Borough Council began negotiating with private sector developers, leading to the publication of a Draft Town Centre Map in 1965. This in turn was soon overtaken by *Teesplan*, which considerably increased the proposed scale of redevelopment following central Middlesbrough's allocated role as a sub-regional centre. In *Teesplan* estimates were made of demand for central area land for civic, commercial and retail uses, based on forecasts of population, income and service sector employment. The policy of concentrating retail provision in central Middlesbrough was problematic because Teesside had two historically established shopping centres, Middlesbrough and Stockton. Neither offered a particularly good range or quality of services, however. Concentrating future growth in Middlesbrough seemed the most effective way of deploying public expenditure to stimulate an improved quality and range of retail services. New office developments were also to be concentrated there. As a result, a considerable share of the forecast increase of 52 000 to 70 000 jobs in service sector activities on Teesside would be concentrated in central Middlesbrough.

Implementing policies to achieve these goals was, however, extremely problematic as successive local authorities lacked the necessary powers and resources. Redevelopment commenced in 1968, and the difficulties that arose and led to its diverging from its intended path are now considered in turn. The first was that they failed to gain widespread support among local residents (Gladstone, 1976). Some residents, dissatisfied with the representation of their interests by local councillors, formed residents associations, which contested the plans and publicly challenged the local authority by proposing 'alternative' plans (Lynch, 1976). As in the case of industrial land policies, environmental concerns triggered an activism that was at variance with the political tradition of passive acquiescence that characterizes much of Teesside. However, these pressure groups essentially modified and delayed, rather than substantially altered, local authorities' planning decisions.

There was more serious opposition from the political and economic interests that would be adversely affected by concentrating development on central Middlesbrough. Even as work began on *Teesplan*, there were proposals to redevelop Stockton town centre, using a joint local authority/private sector partnership (Gladstone, 1976, p. 52). For the local authority, this held out the promise of prestige and a higher rate income; for the company and the big retailing chains, it held out the prospect of handsome profits. It was only with some difficulty that agreement was reached to focus commercial redevelopment on Middlesbrough to the detriment of Stockton.[11] Other sources of opposition from retailing capitals were more easily contained. Local small shopkeepers had limited bargaining power either to oppose proposals or amend them to include their shops in the scheme (Etherington, 1982, p. 89).

A further problem was posed by the reluctance of the major retailing chains to move into Middlesbrough town centre. This was only partially overcome once the local authority indicated that it would veto the development of 'out of town' superstores (*Evening Gazette*, 7 August 1972). By the late 1970s, however, it could no longer sustain this position. The ASDA chain proposed building a superstore at

Eston. Cleveland County Council initially refused planning permission because it would run counter to its Structure Plan proposals. A powerful lobby of local residents and politicians emerged, arguing that Eston had poor retail provision and high unemployment, both of which would be alleviated by the ASDA development. After a five-week public inquiry the Inspector overruled the county council's objections (*Evening Gazette*, 16 November 1978). This decision was all the more serious for the county council's policies because retailing was already being affected by recession and mounting unemployment.

The promotion of Middlesbrough as a sub-regional centre was also undermined by changes in central government policy. The possibility of relocating a major central government office to Middlesbrough had been floated in *Teesplan*. It was not until 1976, after considerable lobbying, that it was decided to relocate the Property Services Agency (PSA), with 3000 jobs, to the town as part of the more general policy of moving government offices out of London. A 13-acre site was acquired and cleared, reputedly at a cost of £1.0 million. However, after the 1979 general election, the Conservative government reversed the relocation decision, providing 'a severe set back to the development of Middlesbrough as a sub-regional centre. It meant that another large area of vacant land faced an uncertain future, adding to the already considerable acreage of vacant land in and around Middlesbrough' (Cleveland County Council Planning Department, 1980, 11).[12]

Implementation of local authority redevelopment policies in central Middlesbrough effectively ended in 1982, a decade after the completion of the first major phase, the Cleveland Centre, which provided 350 000 sq. ft. of retail floorspace and 60 000 sq. ft. of office floorspace.[13] These policies clearly have not been implemented as smoothly or as successfully as originally had been hoped (Cleveland County Council Planning Department, 1980). Middlesbrough's intended role as a cultural and entertainment centre has not materialized. There has been rather more success in attracting major retail stores, but the range and quality of retail services until recently was not considered to be significantly better than in Stockton or Hartlepool. However, with the 1982 opening of the Hill Street Centre, Middlesbrough's share of retail trade rose at Stockton's expense, but there are indications that this successful cementing of Middlesbrough's sub-regional role caused political problems and conflict within Teesside about the location of new retailing facilities (Cleveland County Council Planning Department, 1983b, p. 34). The greatest success has perhaps been achieved in relation to encouraging office development – despite the PSA decision. Even so, supply considerably exceeded demand. By the end of 1983 almost 200 000 sq. ft. of office space was standing empty in central Middlesbrough, while only about 15 000 sq. ft. of additional office space is taken up *annually* in the *whole* of Cleveland County (Cleveland County Council, 1985a, p. 8).

A consequence of this partial and uneven realization of the intended transformation of Middlesbrough's central area is that the associated diversification and growth of employment opportunities has not occurred as planned (Cleveland County Council Planning Department, 1978, p. 35). By the mid-1980s, service sector employment growth was only two-thirds of that assumed in *Teesplan*, and many of these jobs were poorly paid and were part-time (see Beynon *et al.*, 1985). The recognition by the county council that service sector employment had not grown as assumed in *Teesplan* and subsequent Structure Planning documents was a significant one. For it both raised – albeit implicitly – questions on why this should be so and what might be done to combat unemployment on Teesside, as it became clear that manufacturing contraction was in no way compensated for by service sector growth.

5 Conclusions

Around the end of the 1970s and the start of the 1980s a growing number of trades unionists, local authority planners and politicians became aware that the modernization policies of the 1960s had failed and were increasingly the proximate cause of Teesside's problems rather than the solution to them (for example, see Cleveland County Council Planning Department, 1979a, 1980).

The major challenge to modernization policies and the hegemonic project of the 1960s – which some would see as a belated challenge (for example, North East Area Study, 1975) – initially emerged when environmental groups began to organize and protest against them. A growing recognition of their negative environmental impacts combined with the experience of limited employment growth (and then decline) associated with chemicals, oil and steel developments. The county council itself began publicly to question the appropriateness of its past policies. In fact, doubts had been expressed in the early 1970s regarding the compatibility of encouraging further growth in such industries with other local authority policy objectives. For example, Teesside County Council (1972, p. 10) states:

> the image of Teesside as a growth area with an improving environment and widening job opportunities could be adversely affected by the persistence of a labour market dominated by chemicals and steel and an environment under continuous threat from pollution resulting from the development of large capital intensive plants.

But the tone was conjectural and, in the context of the breadth and strength of the alliance favouring further chemicals, oil and steel expansion, at first these reservations were set aside. By the end of the 1970s, however, it was clear that persistent environmental problems and rapid growth in unemployment had been associated with very high levels of fixed capital investment in these activities. Moreover, this had been substantially underwritten by central government regional policy expenditures. Although growing unemployment was extremely serious for Teesside, the fact that chemical and steel producers emerged from this state-sponsored round of restructuring as less than fully competitive internationally also had profound local and national implications. What had seemed in the 1960s to be the great advantages conferred by Teesside's unique location within corporate, national and regional modernization policies, now rendered it uniquely vulnerable to changes in state policies and in the international division of labour in the later 1970s and 1980s. The vulnerability of chemicals, oil and steel to international competition continued to result in serious cuts in capacity and jobs on Teesside and the prospect of further increases in unemployment (Cleveland County Council, 1985a; 1985b). In the 1980s, the county council began publicly to raise these issues and campaign to prevent further job cuts.

As the county council's own assessments of the situation made clear, the unemployment consequences of these cuts for Teesside could have at least been ameliorated if its other policies to attract jobs and diversify employment had succeeded. But its attempts to increase other forms of manufacturing employment by industrial estate development were at best partially and temporarily successful. Although some new jobs were created – hardly a surprise, given the state incentives and the massive pool of surplus labour – they were dwarfed by losses in chemicals, oil and steel and other branches of manufacturing. Total manufacturing employment fell by 15 000 between 1974 and 1977, by 24 000 (or 24 per cent) between 1978 and 1982 (of which 17 000 was in chemicals and steel) (Cleveland County Council Planning Department, 1983a, p. 21). Policies to create new service sector employment were, at best, partially successful. Having risen between 1974 and 1978, service sector employment then fell by 6000 (or 6 per cent) in the next four years, not least because of the multiplier effects of job losses in manufacturing and construction (the latter fell by 10 000, or 41 per cent, between 1978 and 1982 as major construction projects ended).[14] Cleveland County Council soon found its attempt to implement 'autonomous' industrial development and employment policies were necessitated yet also undermined by decisions of central government, nationalized industries and private capital that were beyond its control. In this sense, the hegemonic project of the 1960s was revealed as unrealizable.

Such changes as did occur were associated with a class recomposition, the restructuring of labour markets and growing social polarization within Teesside. In aggregate, relatively well-paid full-time manufacturing jobs for men declined. Conversely, there was some growth in temporary contract work for men within and outside Teesside and an expansion of poorly paid, part-time (often sub-contracted) service sector work, which predominantly went to women. But perhaps the starkest indicator of the changes that occurred was the rise to more than 20 per cent in Cleveland County's registered unemployment rate. Towards the end of 1983 Cleveland had the unenviable distinction of being the county with highest registered unemployment rate in Great Britain (Cleveland County Council, 1985b, para. 3.4). Government temporary employment schemes became one of Cleveland's biggest employers. Within Teesside, the unemployed were concentrated in the old working-class housing areas associated with declining industries and in local authority sink estates such as East Middlesbrough. Poverty grew in the increasing numbers of households there with no employed member. Elsewhere in Teesside, however, there was comparative affluence: for example, in the new owner-occupied housing estates of Coulby Newham, towards the rural periphery. This was especially so in two-income households where married women had become wage workers. Thus, industrial decline was linked to growing social and spatial polarization.

Because of the worsening unemployment situation, the county council successfully pressed for Cleveland to get Special Development Area status in June 1982 and for part of the old Ironmasters district of Middlesbrough to become an Enterprize Zone in November 1982. It is perhaps indicative of growing central government recognition of Teesside's unemployment problems, as well as of effective lobbying by and on behalf of Teesside interests, that this was the *only* case of upgrading to SDA status then announced. The same point can be made with regard to the establishment of an Urban Development Corporation on Teesside in 1987. Whether such initiatives will significantly contribute to easing unemployment or reversing social polarization remains to be seen, but the prognosis has to be a poor one. The problem for local authorities is to devise new politically acceptable policy responses in a situation where their own powers and resources are extremely limited. Most fundamentally, however, the problem for them is that the key decisions affecting Teesside lie in the hands of the UK government, BSC and ICI, and will continue to be more and more shaped by the pressures of international markets, as the county council increasingly acknowledges (Cleveland County Council, 1985a, pp. 18–19). In these circumstances, although the hegemonic modernization project of the 1960s has been blown apart, there is no locally based replacement in prospect. Indeed, the very notion of such a possibility has been effectively discredited. Rather, it is widely accepted that the fate of the locality must be to dance to the tune of international markets, mediated via government policies and centrally appointed agencies such as the UDC, leaving the Labour-controlled local authorities increasingly to cope with the consequences of mass unemployment (Cleveland County Council, 1987).

Acknowledgement

This paper is a revised (December 1988) version of Hudson, R. 1988. *Middlesbrough Locality Study, Working Paper* 7. It draws on research carried out with Huw Beynon, Jim Lewis, David Sadler and Alan Townsend. The usual disclaimers apply, however.

Notes: Chapter 3

1 This section draws on Etherington (1982).
2 Before 1968, the Local Planning Authorities were Hartlepool and Middlesbrough County Boroughs and Durham and North Riding County Councils. Stockton Municipal Borough had 'delegated authority'. In addition, Redcar Municipal Borough, Billingham, Eston and Thornaby Urban District Councils and Stockton Rural District Council all had an interest in this development.
3 The process was further complicated because, following the creation of Teesside County Borough, the former *Teesplan* area was divided between it, Durham County and the North Riding of Yorkshire. Consequently, three local authorities were responsible for preparing Structure Plans for the area covered by *Teesplan*.
4 Indeed, the links between these regional planning documents and planning documents relating specifically to Teesside were extensive. For example, the analyses in *Challenge of the Changing North* were used as the basis for preparing the sub-regional employment and population forecasts on which *Teesplan's* proposals were based. This, of course, represented one way of attempting to achieve some consistency between planning at these different spatial scales.
5 The parallels with the representation and prioritization of the National Coal Board's interests within the coalfield areas of County Durham are striking. Such processes are by no means confined to North-East England, however (Kunzmann, 1982).
6 This in many ways echoed events on the Durham coalfield. Thus, for example, while *Teesplan* assumed declining employment in chemicals and steel, it assumed that 53 000 additional jobs in other sectors of manufacturing (75 per cent of them for men) should and would be

attracted from outside Teesside in the period to 1991 to help fill the 'job gap' of 61 000 that would occur in this period (Wilson *et al.*, 1969, vol. 1, pp. 20–7). Ignoring for the moment the serious methodological flaws in the 'job gap' approach (Regional Policy Research Unit, 1976, pp. 39–70), the point here is that this *assumed* growth in employment then became a central element around which the rest of *Teesplan's* proposals were built.

7 For a technical critique of these policies, see Regional Policy Research Unit, 1976, pp. 39–70; also North East Area Study, 1975, pp. 92–102.

8 In fact, the first new plant was a joint Phillips–ICI oil refinery. It was also significant because, for the first time, ICI was in a position to obtain some of its feedstocks from a refinery on Teesside. Until 1966, its Wilton works – uniquely among major petrochemical complexes – had been reliant upon importing refined feedstocks (Hudson, 1982).

9 One reason why the 'jobs not birds' lobby was such a powerful one was the impact of BSC policies in causing cuts in steel industry employment in Hartlepool (see Hudson and Sadler, 1986).

10 This 'job gap' methodology is an extremely dubious one (see note 7). The point here, however, is not its inadequacies but rather the implications of its use for subsequent land provision policies.

11 This competition *between* local authorities on Teesside regarding their place in the service sector hierarchy was a recurrent one. For example, when BSC sharply cut its employment at Hartlepool, Ted Leadbitter, West Hartlepool's Member of Parliament, argued that alternative office employment should be directed to Hartlepool and so away from central Middlesbrough.

12 It stood empty for several years, as an open-air car park (presumably one of the North-East's more expensive ones). Its subsequent development was a result of the activities of the Urban Development Corporation.

13 This was part of the 400 000 sq. ft. of office space *claimed* to be built or under construction in central Middlesbrough between 1966 and 1970 (Teesside Borough County Council, 1973, p. 32).

14 These actual job losses may be compared with Structure Plan forecasts of growth, 1971 to 1991:

chemicals	+ 4800
steel	+ 4300
services	+ 39 000

The differences are a cutting comment on the character of 'planning' in the context of capitalism in the UK.

References: Chapter 3

Bell, Lady. 1907 (reprinted 1985). *At the Works*. London: Virago.

Beynon, D., Hudson, R., Lewis, J., Sadler, D. and Townsend, A. R. 1985. 'Middlesbrough: contradictions in economic and cultural change'. *Middlesbrough Locality Study, Working Paper No. 1*. Durham: University of Durham.

Beynon, D., Hudson, R. and Sadler, D. 1986. 'The growth and internationalization of Teesside's chemicals industry'. *Middlesbrough Locality Study, Working Paper No. 3*. Durham: University of Durham.

Beynon, H., Hudson, R. and Sadler, D. 1989. *A tale of two industries: one coal mine, one steelworks and the destruction of a region's economy*. Milton Keynes: Open University Press.

Board of Trade. 1963. *The north east: a programme for regional development and growth*. Cmnd 2206 (The Hailsham White Paper). London: HMSO.

Briggs, A. 1963. *Victorian Cities*. Harmondsworth: Penguin.

Carney, J. and Hudson, R. 1974. 'Ideology, public policy and underdevelopment in north east England.' *North East Area Study Working Paper No. 6*. Durham: University of Durham.

Cleveland County Council. 1977. *Teesside Structure Plan*. Middlesbrough: Cleveland CC.

Cleveland County Council. 1979. *Report of the Panel: Appendix 2, Miscellaneous documents*. Middlesbrough: Cleveland CC.

Cleveland County Council. 1980. *The case for Cleveland County*. A submission made to the Secretary of State for Trade and Industry. Middlesbrough: Cleveland CC.

Cleveland County Council. 1985a. *Cleveland Review, 1974–84*. Middlesbrough: Cleveland CC.

Cleveland County Council. 1985b. *Crisis in Cleveland (from dream to nightmare)*. Middlesbrough: Cleveland CC.

Cleveland County Council. 1987. *Unemployment strategy*. Middlesbrough: Cleveland CC.

Cleveland County Council Planning Department. 1975. *The economic impact of North Sea oil in Cleveland*. Middlesbrough: Cleveland CC.

Cleveland County Council Planning Department. 1978. *Industrial structure and employment trends and their consequences for unemployment in Cleveland*. Middlesbrough: Cleveland CC.

Cleveland County Council Planning Department. 1979a. *Employment, Discussion Paper No. 1*, County Structure Plan Reassessment Programme, County Planning Department Report No. 163. Middlesbrough: Cleveland CC.

Cleveland County Council Planning Department. 1979b. *Annual Monitoring Report*. Middlesbrough: Cleveland CC.

Cleveland County Council Planning Department. 1980. *Sub-regional centre and shopping, Discussion Paper No. 3*, County Structure Plan Reassessment Programme. County Planning Department Report No. 170. Middlesbrough: Cleveland CC.

Cleveland County Council Planning Department. 1983a. *Employment Review*. Planning Department Report No. 232. Middlesbrough: Cleveland CC.

Cleveland County Council Planning Department. 1983b. *Cleveland Review, 1983*. Planning Department Report No. 237. Middlesbrough: Cleveland CC.

Etherington, D. 1982. *Local authority policies, industrial restructuring and the unemployment crisis: an evaluation of the formation and impacts of local economic initiatives in Cleveland, 1963–82*. Unpublished MA thesis, University of Durham.

Firn, J. 1975. 'External control and regional development: the case of Scotland.' Environment and Planning A, **VII**, 393–414.

Gallant, V. 1982. 'Economic and employment initiatives in Cleveland.' *Northern Economic Review*, **7**, 11–17.

Gladstone, F. 1976. 'Teesside: sprawl gone bad'. In F. Gladstone (ed.), *The Politics of Planning*. London: Temple Smith.

Gregory, R. 1976. 'The Cow Green Reservoir'. In P. J. Smith (ed.), *The politics of physical resources*. Harmondsworth: Penguin.

Hamer, M. 1974. *Wheels within wheels: a study of the road lobby*. London: Friends of the Earth.

Hudson, R. 1982. 'Capital accumulation and chemicals production in Europe in the post-war period.' *Environment and Planning A*, **15**, 105–22.

Hudson, R. 1983. 'Capital accumulation and regional change: a study of north east England, 1945–83', pp. 75–101. In F. E. I. Hamilton and G. Linge (eds), *Regional industrial systems*. Chichester: Wiley.

Hudson, R. and Plum, V. 1986. 'Deconcentration or decentralization? Local government and the possibilities for local control of local economies.' In M. Goldsmith and S. Villadsen (eds), *Urban political theory and the management of fiscal stress*, pp. 136–60. Farnborough: Gower.

Hudson, R. and Sadler, D. 1985. 'The development of Middlesbrough's iron and steel industry, 1841–1945.' *Middlesbrough Locality Study, Working Paper No. 2*, Durham: University of Durham.

Hudson, R. and Sadler, D. 1986. 'Contesting works closures in Western Europe's old industrial regions: defending place or betraying class?' In A. J. Scott and M. Storper (eds), *Production, work, territory*, pp. 172–94. London: Allen and Unwin.

Kunzman, K. R. 1982. 'Structural problems of an old industrial area: the case of the Ruhrgebeit.' Dortmund: Institut für Raumplanung, Universitat Dortmund (mimeo).

Lynch, D. 1976.'Planning and participating democracy in a district of Middlesbrough.' *Working Papers in Social Anthropology, No. 2*. Durham: University of Durham.

Marriott, O. 1967. *The Property Boom*. London: Pan Books.

National Ports Council. 1972. *Annual Report, 1972*. London: HMSO.

North, G. A. 1975. *Teesside's economic heritage*. Middlesbrough: Cleveland County Council.

North East Area Study. 1975. *Social consequences and implications of the Teesside Structure Plan*, Vol. 1. Durham: University of Durham.

Northern Economic Planning Council. 1966. *Challenge of the changing North*. Newcastle: NEPC.

Northern Economic Planning Council. 1969. *Regional ports survey*. London: HMSO.

Pettigrew, W. 1985. *The awakening giant: continuity and change in ICI*. Oxford: Blackwell.

Priestley, J. B. 1934. *English Journey*. London: Heinemann.

Regional Policy Research Unit. 1976. *A preliminary report on current industry and employment analyses, forecasts and strategies for the North East: a study of regional and county planning documents*. Aycliffe and Peterlee Development Corporation, Lee House, Yoden Way, Peterlee, County Durham.

Regional Policy Research Unit. 1979. *State policies and regional uneven development: the case of North East England*, Part 9. Final Report No. RP 270. London: Centre for Environmental Studies.

Robinson, J. F. F. and Storey, D. 1981. 'Employment change in manufacturing industry in Cleveland, 1965–76.' *Regional Studies*, **15** (3), 161–72.

Teesside County Borough Council. 1972. *Teesside Structure Plan: report of survey, 1972*. Middlesbrough: Teesside CBC.

Teesside County Borough Council. 1973. *Teesside Structure Plan: draft document for participation and consultation*. Middlesbrough: Teesside CBC.

Wilson, H. and Womersley, L. and Scott Wilson Kirkpatrick and Partners. 1964. *Teesside Survey and Plan: Project Report*. London: HMSO.

Wilson, H. and Womersley, L. and Scott Wilson Kirkpatrick and Partners. 1969. *Teesside Survey and Plan*, Vols. 1 and 2 (also known as *Teesplan*). London: HMSO.

Working Group on the Implications of North Sea Oil for the Development of the North Tees Area. 1972. *Report*. Middlesbrough: Teesside County Borough Council.

4

Merseyside in crisis and in conflict

RICHARD MEEGAN

1 Introduction

By the late 1960s, about one in eight of Merseyside's population lived in the eastern crescent of predominantly local authority housing estates (Merseyside Council for Voluntary Services, 1978). It was these outer estates on which our research focused and particularly, because of their size and importance as industrial centres, on Kirkby in the north and Halewood and Speke in the south. As Figure 4.1 shows, these estates fall within the remit of two local authorities: Halewood and Kirkby are in Knowsley Metropolitan Borough and Speke is in the City of Liverpool. They thus offered the possibility of exploring local variations in political and policy response to economic restructuring. All three estates are in the same local labour market (the Liverpool 'Travel-to-Work-Area'), so analysis of labour market changes was framed accordingly. Political and policy response, however, required analysis in a different spatial context (for example, local authority areas for statutory responses and neighbourhoods for community-based initiatives).

New industries had located in these outer areas, notably in the late 1950s and early 1960s when the branch plants of large national and multinational corporations moved in and expanded to take advantage of the available labour force and regional policy incentives. Given the scale of this investment, it was understandable that commentators were reasonably optimistic about the region's economic growth potential (Lloyd, 1970; Smith, 1969). But, alas, it was not to be. By the mid 1980s, Liverpool had slipped down to sixth place in the league table of Britain's ports, as its location was devalued by the nation's economic and geopolitical turn towards European trade and political integration (Lane, 1987). Between 1971 and the mid-1980s, around 20 000 people employed in dock work and sea transport lost their jobs, a 75 per cent decline (Meegan, 1988a). The newer manufacturing sectors followed suit, with Liverpool losing more than half of its manufacturing jobs in the same period. About 95 000 jobs disappeared and, with their relatively heavy dependence on manufacturing, the outer areas were particularly hard hit.

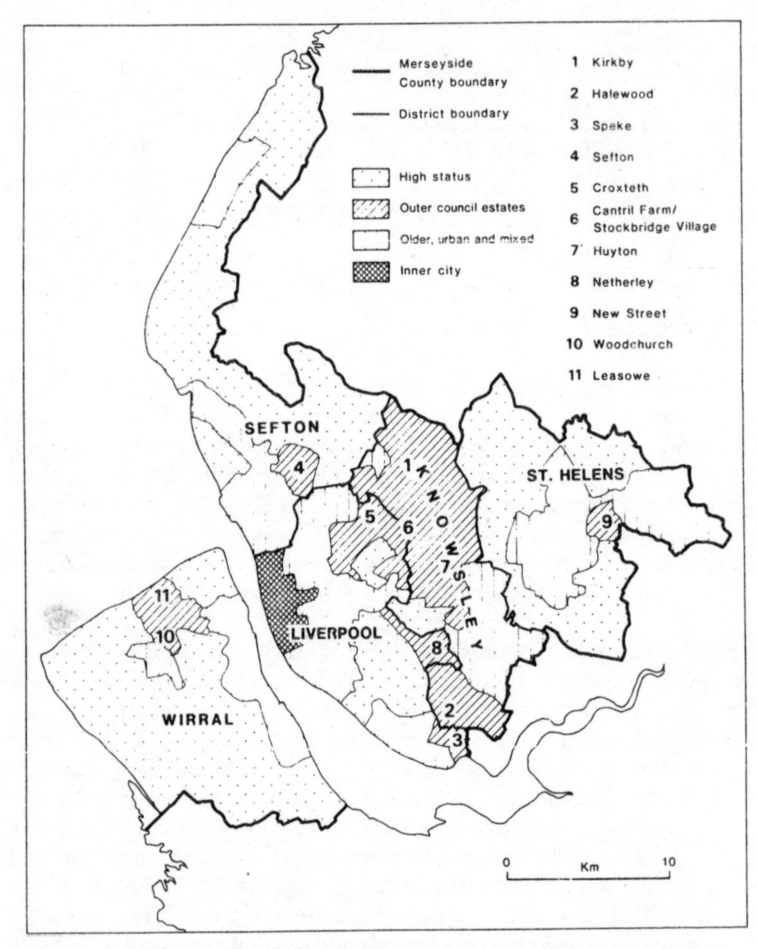

Figure 4.1 Merseyside's social areas including the major outer estates.

Registered unemployment is on average double the national rate and, in parts of the estates, such as New Hutte in Halewood and Tower Hill in Kirkby, more than treble. Poverty is endemic. On an admittedly crude measure of 'living standards' derived mainly from 1981 census data, Knowsley – with its high concentration of outer estates – is ranked last of 457 local authority districts, separated from Liverpool only by Easington (CES Ltd, 1988).

This growing impoverishment is one harrowing indication of Merseyside's demotion in the national and international division of labour. Another is the scale and pace of population decline. In the last decade, Merseyside became the fastest declining metropolitan county. Recent estimates show this continuing into the next century, with Liverpool and Knowsley being the two fastest declining English metropolitan districts (Office of Population Censuses and Surveys, 1988).

Two Labour-controlled local authorities were faced with the task of coping with this decline in the 1980s (Knowsley since its inception with local government reorganization in 1974 and Liverpool since 1983). The councils had to contend with a Conservative government that began its second term with a massive majority and a radical manifesto that included the measures aimed at restricting local government autonomy and expenditure outlined in Chapter 1.

In the seven years before the election of the Labour administration in 1983, rate support grant in Liverpool fell from 40 to 29 per cent of the City's total revenue and the council estimated that around 70 per cent of the 1979–83 rate increase resulted from this grant reduction (Liverpool City Council, 1983). The Knowsley rate base came under similar pressure. As rate support declined from 46 per cent of total income in 1979 to 29 per cent in 1986, rates were squeezed up from 5 to 11 per cent of the total in an attempt to maintain services.

Difficult political choices faced both authorities. However, the strategies adopted were very different. In Knowsley, Labour opted to remain within the law, to set 'legal budgets' and, albeit begrudgingly, to cooperate with central government. As will be argued later, this pragmatism had its political costs, facing the council with policy choices that exacerbated local political differences.

A very different political strategy was adopted by the newly elected Liverpool Labour administration in 1983; a strategy on which the House of Lords eventually passed judgement in Spring 1987, upholding the High Court's verdict that a group of Labour councillors were guilty of wilful misconduct in delaying the setting of a rate. Forty-seven councillors were disqualified from office and ordered to pay surcharges and costs of £348 000. Two councillors – Derek Hatton, the deputy leader, and Tony Mulhearn – had also been expelled from the Labour Party for their connections with 'the Militant Tendency' (a third councillor, Felicity Dowling, was expelled on these grounds in 1987).

The different political stances of the two councils were reflected in policy development. In both cases housing was the dominant issue. Housing – especially public housing – has long dominated politics on Merseyside. Liverpool was the first city to build public housing, in 1869, and Liverpool and Knowsley have among the highest concentrations of public housing in Britain and some of the worst housing conditions (CES Ltd, 1988).

The most single-minded and, as it turned out, the most politically provocative initiative was Liverpool City Council's ambitious council-housing-based 'Urban Regeneration Strategy'. This strategy was given priority over all other policies including local economic development. Its funding lay at the heart of the council's stand against government financial restrictions and eventually led to the expulsion of most of the Labour council from office. In contrast, reflecting its general political stance, Knowsley Council's housing policy was based on greater collaboration with government and the private sector. Relatively orthodox initiatives were also pursued in other policy areas such as economic development, albeit rather belatedly and with restricted resources.

As is argued elsewhere in this book, the nature and 'effectivity' of these political and policy responses were structured by a complex combination of local conditions – economic, political and institutional – interacting with national and international processes of economic and political change. On the economic front, providing the *leit motif* for the overall political and policy response, there is the catastrophic decline of the regional economy and its subordination in the national and international economy. This has constrained local responses in two ways. First, through the decline in income and local resources already noted; secondly, as will be discussed later with reference to Knowsley, through the socio-economic legacy of the region's previous participation in the spatial division of labour.

The effectiveness of both local authorities' strategies was hindered by these severe resource constraints, but it was also affected by the political context, a context that the strategies themselves also partly created. Thus Liverpool was not just handicapped by declining government financial support but also by the political opposition from government that its strategy provoked. This was compounded by significant local opposition. Even in Knowsley, policies ran up against government spending restraints and choices were made that, combined with existing local political differences, contributed to internal conflict that pushed Knowsley on to the national political stage.

2 Economic restructuring and policy response: the legacy of history

The restructuring of the 1970s and early 1980s did not just mean devastating job loss, it also signified the end of an important transformation of the local labour market. The implications of this can be seen particularly clearly in Knowsley, where the outer estates housed the 'overspill' population of inner Merseyside, an important labour pool for the post-war manufacturing investment in the region. This mainly required semi-skilled and unskilled workers, and the occupational and skill profile of the local workforce was shaped accordingly (Meegan, 1988b).

Adjustment to this new production regime had been difficult for a workforce steeped in the traditions of casualism – especially men working in dock and dock-related work – or which only had limited previous experience of work – the women who formed the core workforce for the new food-processing and light engineering factories. The high levels of absenteeism in the early years of the industrial estates reflected these difficulties. For those who had worked before, a degree of job control was exchanged for the prospect of secure employment and a regular pay packet. However, the labour shakeout in the restructuring of the 1970s and early 1980s ended this compromise. Mass unemployment returned to a population with little capital and few marketable skills, as the large plants closed or radically transformed their production with labour-saving investment in new technologies.

In these circumstances local politicians and policy-makers are left with an apparently intractable task: how to regenerate a 'local economy' with depressed local income and demand and a tightly circumscribed skill base. The rationale of regeneration being founded on a locally based 'enterprise culture', the ideological model championed by the current Conservative government, is called into question. The 1981 Census shows that, of all local authority districts in the North-West, Knowsley has the lowest proportion of self-employed people, and more recent surveys confirm this picture. Thus, Knowsley had by far the lowest number of 'starts' on the Enterprise Allowance Scheme in the mid-1980s. A local economic development officer also made a telling comparison between small firm formation in Knowsley and Crewe, areas of similar size. In a two-year period, Crewe had seen the start-up of 200 businesses. Knowsley had taken four years to produce 140 (although the survival rates were almost identical).

For some, the failure of an 'enterprise culture' project reflects the personal failings of local people and the 'locality':

> 'Of course, the trouble with your constituents is that they're not self-starters, are they? They don't start up their own businesses. They've no entrepreneurial spirit. They've got no get-up-and-go.'
>
> (The Prime Minister, Margaret Thatcher, describing the constituents of Knowsley North to its then Member of Parliament, Robert Kilroy-Silk. Quoted in Kilroy-Silk, 1985, p. 45)

But this interpretation is blind to the historical evolution of the local labour market and the structures shaping individual choice. What skills did standardized mass-production jobs develop and what did the labour processes do to develop self-confidence or 'get-up-and-go'?

The nature of local capital was also shaped by the process of labour market transformation. This also had implications for the local response to the recent restructuring. Post-war investment resulted in branch-plant factories and production and labour processes dominated by multinational corporations. As a local economic development worker stated:

> There's no atmosphere of a business community in Knowsley . . . because the Head Offices of the big companies are not located here. They're all branch plants. You don't get any sense of an integrated company atmosphere. (Interview 1987)

This branch plant economy has inhibited the development of the local economic base. The use by the large corporations of their own (often transnational) supply lines has limited local supply networks. Ironically, this neglect may now be coming home to roost – if the arguments about the superior competitiveness of spatially concentrated supply systems are correct.

Inter-linkages are also important for producer services, one of the leading growth areas in the economy (Daniels, 1986). In Knowsley, as the economic development officer acknowledged, these links are also lacking:

> there is one firm of accountants in Knowsley. One. It's incredible. And he's in Prescot, operating practically as a one-man band! Yet there are 250 small firms on Knowsley Industrial Park looking for accountants. They have to go to Liverpool. There are 10 small firms of accountants in Warrington (New Town) Centre alone. There's one accountant in Knowsley, but 69 turf accountants! It's just a classic example of the unbalanced social structure of Knowsley. (Interview, 1987)

This dearth of local accountants partly reflects Knowsley's social structure but, again, to avoid slipping into 'blaming the victims', what about the role of the industrial structure and the previous spatial division of labour, in influencing this social outcome? With the large corporations relying on their own internal accounting structures, there was no incentive for the local development of this producer service.

3 Economic restructuring and the radicalization of local politics

Liverpool's late arrival in the Labour fold (its first Labour council was elected in 1955 and its parliamentary seats were dominated by

Conservatives until the early 1960s) always marked it out from other large Northern cities, as did its ten-year interlude with a Conservative-backed Liberal administration after 1973. The early weakness of the Labour Party was influenced by the sectarian nature of local politics (undermined, it should be noted, by the mixing that occurred when population was dispersed to the outer estates – a progressive but unintended effect of this social engineering). The more disastrous aspects of the overspill programme – the high-rise building, initial social isolation and the break-up of inner-city communities – also influenced political development, engendering disillusionment with Labour, which had enthusiastically pursued population dispersal, making housing a key political issue and laying the ground for Liberal 'community-based' politics in the 1970s.

Yet there has been a longstanding radical socialist presence on Merseyside. Indeed, the Labour Party faction that was so influential in 1983–7 originated in this region (Taafe and Mulhearn, 1988). And it was Liverpool that put the Militant Tendency firmly on the national political map through the Militant-inspired Labour City Council of this period. Given the acceleration of economic decline in the region during the 1970s, it is not difficult to understand why such a political stance emerged.

Selective emigration had reduced the middle-class vote and reinforced Liverpool's position as England's most proletarian city. In 1981 the city had the lowest proportion of employers, managers and professional people in its active economic population – 9 per cent compared with an urban average of 14 per cent (Liverpool City Council, 1984). Between 1971 and 1981 there was a significant increase in the number of workers unable to enter any employment status on their census returns. These 'unclassified' workers (mostly unemployed and many who had never worked) nearly doubled their share of the economically active population – from 4.4 to 8.5 per cent between 1971 and 1981. Taken with the semi-skilled and unskilled workers, the unskilled and/or unwaged working class made up 41.3 per cent of the city's economically active population, compared with an average for English cities of 30.3 per cent.

As Urry (1981) argues, the growing incorporation of localities in larger-scale restructuring processes raises the potential for increased politicization. This certainly appears to have happened on Merseyside. The speed and pervasiveness of the restructuring of industries that two decades earlier were heralded as the saviours of the area – and the vulnerability of Merseyside's role as an outpost of transnational corporations – provoked frustration and cynicism among workers (Meegan, 1988a). This resulted in an openness to more confrontational politics or at least to those that 'put the interests of Merseyside and Merseysiders first'. As one activist in a Speke community group put it:

> If there is a cut to be made, Liverpool gets it. That's why Hatton [the deputy leader of Liverpool City Council and a leading 'Militant supporter'] and company were popular – because he fought Thatcher. The only way people here ever got anything was to fight for it. Our union supported the Council when they asked us to come out in support of the stand against the government. It may sound strange, that, because twelve months later our action group opposed them. (Interview, 1987).

The Labour Party offered a radical response to the decline, its emphasis on municipal housing-led regeneration matching public despair at the deterioration in the city's built environment. This policy had been developed by the Left, its main architect was Tony Byrne – avowedly 'non-Militant' and later chair of the all-powerful Housing and Finance Committees – and was eventually adopted by the District Labour Party. This important policy-making body played a crucial role: while Militant supporters were in a minority on the council, they dominated the District Labour Party, which allowed them to steer council policy development (Crick, 1986).

The Urban Regeneration Strategy involved extensive renovation and rebuilding of the City's council housing, focusing on twenty-two Priority Areas (containing the worst housing and social conditions). The aim was to provide houses with gardens surrounded by defensible space. The Strategy had priority over all other policies. This embodied an important political statement. The council believed that its ability to influence the private sector (to which it was anyway ideologically opposed) was limited. Tony Byrne stated 'I can't do anything to locate a new factory in Speke or anywhere else, but what we can do is to deal with unemployment, and the environment and living conditions within the limits of our capabilities' (Andrews, 1985; see also Liverpool City Council, 1986, Appendix I).

The Council thus did not share the growing local authority enthusiasm for local economic development agencies. It had a resolutely municipal and, in many ways, traditional approach. Economic development was based on a programme of municipal employment, an attempt to reverse the rundown of public sector employment in the previous decade, and on the Urban Regeneration Strategy, through employment generated by the housebuilding programme and by the new municipal leisure facilities that formed part of the strategy.

Housing also loomed large in the policies of Knowsley, which, following local government reorganization in 1974, included many of the largest Liverpool overspill housing estates such as Kirkby and Halewood. As noted above, this Labour stronghold opted for a collaborative policy stance. However, this did not mean that local politics had not been radicalized. Not surprisingly, given its declining economic features, there had been a substantial increase in local politicization. For example, the first union branch for unemployed workers was set up after the closure of one of the major outer estate factories (Standard Triumph in Speke), preparing the ground for a network of unemployed centres across Merseyside. The most active of these is based in Kirkby: in many ways, this centre and its activists have come to symbolize the radicalization of local politics.

In 1971 there were around 30 000 jobs in Kirkby, mainly in its large post-war industrial estate. By the mid-1980s there were less than 19 000, a decline of 38 per cent. Unemployment became *the* political issue, eclipsing the housing problems and youth alienation that had dominated the area's settling-down period in the 1960s. These earlier problems had been important politically, however, because they produced several action groups, at first mainly concerned with housing issues.

Groups formed to lobby local government (first Liverpool City Council, then Kirkby Urban District Council and then Knowsley Metropolitan Borough Council). A major issue was council housing allocation policy, particularly as it applied to the unpopular maisonettes and high-rise flats. Flatdwellers' Associations were formed on each of the original housing estates in Kirkby (Southdene, Northwood and Westvale). These pursued public campaigns for improved housing conditions (bringing in Shelter, for example, to support their case against the tower blocks).

But it was the last Kirkby estate to be built – the 'overspill' of the 'overspill' – Tower Hill, that produced one of the most politicized campaign groups, the Tower Hill Unfair Rents Action Group. Its genesis lies in the 1972 Housing Finance Act, which necessitated council rent rises. After the Kirkby Urban District Council voted to implement the Act, the Group led a rent strike. Within two months about half of the households in Tower Hill were refusing to pay rent. The strike did not end until December 1973. Debts of more than £200 000 had been run up, and nine strikers were jailed. Legal action against tenants was dropped in exchange for an agreement on repayment of arrears and the group disbanded. However, the campaign left an indelible impression on local politics, and its legacy influenced the later response to unemployment – not least because some workers in the unemployed centre had had experience of the earlier rent strike and housing campaigns.

The workers at the Unemployed Centre campaigned not just in support of the unemployed but also for the employed in the area. Campaigns have focused on women 'signing-on', the 'right-to-fuel' and benefit take-up. The Centre also provides an important base for social and cultural activities. But it is the Centre's involvement in local employment issues, through links with the trades unions, that has proved politically controversial. It has involved supporting workers in dispute with local employers in some of the small firms that have located on the industrial estate after the disappearance or contraction of the large employers. This has provoked local media coverage variously – and, from all accounts, mistakenly – casting the Centre as a 'den of Militants' and caricaturing its supposedly 'rent-a-picket' tactics.

This coverage was taken up and magnified by the national media in relation to the reselection of Kilroy-Silk, the sitting MP for Knowsley North, in 1986–7. Kilroy-Silk's resignation and the Labour Party National Executive Committee's decision to impose a candidate against the wishes of the local party provoked a further round of critical media coverage. Traumatized by the loss of two general elections, and with one pending, the Labour Party leadership was determined to present what it saw as an electable, 'realist' image, suppressing 'local difficulties' such as the selection of an 'unsuitable' parliamentary candidate (in Knowsley) or an apparently 'Militant-inspired' confrontation with central government (in Liverpool).

But, as we have noted, in contrast with Liverpool, in Knowsley this radicalization of politics has not been translated into Council policy. There are several reasons for this. As Knowsley has been Labour-controlled since its formation in 1974, there have been no parallels with the Liverpool situation in the 1970s and early 1980s, when Labour was the largest single party but was kept out of office by centre–right coalitions. The political frustrations that such situations engender has been avoided. A close and relatively stable relationship between officers and members has developed, which has tended to favour a professional, 'working the system' approach. This is strengthened by the single-party dominance of the Council. Officers have never 'worked for the other side' and there is no need for political appointments to make sure that the party line is promoted. Policy has developed in a more evolutionary way than in situations where there are frequent swings in political control.

Of course, even within 'one-party' states there is political debate. In Knowsley this debate has been dominated by the centre–right. Knowsley has two parliamentary constituencies, Knowsley North and Knowsley South, and two broadly corresponding divisions within the Labour Party. Knowsley North tends to be more to the left than Knowsley South. Given that Knowsley North is dominated by Kirkby (with its history of opposition to Council policy) this is not surprising. Also, politics in Knowsley South is based around Labour clubs, which, with their social emphasis, are arguably less 'radical' than the more politically based party organization in Knowsley North.

However, Knowsley South councillors outnumber those from Knowsley North. Moreover, until recently, the adversarial stance that Kirkby activists have adopted towards the Council has inhibited them from seeking political office. So the influence of radical left politics has been marginal in Knowsley, and this has been reflected in its political strategy and policies. Policies have included collaboration with government – in 'community refurbishment', housing demolition and conversion, and environmental improvement – and with the private sector, notably in handing over a large run-down housing estate, Cantril Farm (renamed Stockbridge Village), to an independent housing trust made up of representatives of the local authority, a construction company and a building society.

Attempts have also been made to develop a local economic strategy, with the establishment in 1983 of an Economic Development Unit, an Enterprise Agency and the purchase from Liverpool City Council of the Kirkby Industrial Estate (re-named the Knowsley Industrial Park). The council has also backed the establishment of a local Chamber of Commerce, to help create the 'integrated company atmosphere' that previous economic development has singularly failed to provide.

However, these policies have been hindered by government spending restraints and overshadowed by the scale and seriousness of the housing and social needs. The economic development budget is tightly constrained, the Economic Development Unit's current budget is about £60 000. The resources available for economic development cannot hope to match, for example, those available to the council's closest and most successful rival for economic development, Warrington New Town.

4 Policy implementation: the local political dimension, class, race and community

The surcharging and removal from public office of the 47 Liverpool councillors underscored the centre's dominant position in the central–local government relationship. Effective local policymaking depends on the establishment of a *modus vivendi* between the two parties. Where this breaks down, as in Liverpool from 1983–7, no amount of locally contrived 'creative accounting' or independent funding (such as Liverpool's sale of £40 million of mortgages to a French bank or its deferred purchase loans of £70 million from Japanese and Dutch banks) can protect local policies from compromise.

However, it is not just the relationship between central and local government that shapes local policy development and implementation. The state of local politics is crucially important and local political differences take on added significance in areas at odds with central government. Here, the clichéd caution that 'together we stand, divided we fall' becomes more meaningful, as the history of Liverpool City Council's 1983–7 political strategy illustrates. Media coverage of the City Council focused on political divisions between the council and inner city groups and the deterioration in the council's relationship with its own trades unions. However, there were also divisions between the council and groups on its outlying housing estates. These were less publicly debated, but equally indicative of the local undermining of the council's political strategy.

At the time of the election of the 1983–7 Labour administration, the Merseyside housing cooperative movement was concentrated in inner Liverpool. Cooperatives have been a growing force in Liverpool housing (CDS, 1987; Macdonald, 1986), but they have aroused a great deal of controversy, especially on the left. The attitude of the powerful District Labour Party was set out in its 1984 housing policy statement. This viewed them as 'part of a deliberate and calculated attack on municipal housing by the Tory party nationally, aided and abetted by the local Liberal/Tory alliance' (quoted in Grosskurth, 1985). This led to the municipalization of seven housing cooperatives, in varying stages of development, by the new Labour administration.

One scheme had been initiated by a group based in the Vauxhall inner-city area, the Eldonian Residents Association. The 'Eldonians' became one of the council's most implacable opponents. Unwilling to admit defeat, the Eldonians turned their attention to a derelict industrial site and gained direct central government support for a scheme to build the largest family housing cooperative in the country. The council refused planning permission and a protracted and bitter battle followed. This involved a struggle to control the ward Labour Party. As one of the main activists explained: 'We recruited over 100 people overnight . . . and all because of the dogmatism of the Militants. They've created the strongest Labour ward in the city here' (quoted in Grosskurth, 1985; see also *Architects Journal*, 1988).

The cooperative eventually won planning permission on appeal, and embarked on an even more ambitious scheme involving community business projects as well as housing. But this victory had been at the cost of much political in-fighting within the local Labour Party. The divisions remain, although the potential for conflict has now been reduced by an extension of the boundaries of the Merseyside Development Corporation, which now takes in the Eldonian scheme. Whether this was done expressly to achieve this objective is not clear, but the move was generally welcomed by the Eldonians. For the council, the result is an erosion of its authority and a hard lesson in 'community politics'.

Toxteth, or 'Liverpool 8', is an inner-city area that is perhaps more widely known than Vauxhall. In 1981 it was the first area in mainland Britain to experience the police use of CS gas to contain civil disturbance. It contains several active ethnic minority groups, some of whom formed the 'Black Caucus' to promote the interests of Liverpool's ethnic minorities in local politics. This group ran one of the most obstinate and debilitating campaigns against the council. It concerned what was seen as a politically motivated and locally insensitive appointment of a Race Relations Adviser by the council, but was also fuelled by the group's more general hostility towards the council's racial politics (Liverpool Black Caucus, 1986; Ben-Tovim, 1989; for a response, see Taafe and Mulhearn, 1988). The campaign involved public demonstrations, including occupying the office of the deputy leader of the council, Derek Hatton. These demonstrations opened up sharp divisions within local politics, bringing into question the representativeness of the local council, whose whole stance was dependent on undivided support. The impact of the Caucus's actions was heightened by the refusal of one of the council's trade unions, NALGO, to recognize the appointment of the Race Relations officer. The appointee was isolated inside the council administration as well as in the local black community.

The council's trades unions played a key role in the campaign against central government over rate-setting. Deep political schisms occurred as the unions faced a strategy that depended on their accepting the (in the council's view) temporary redundancy of their members. The consequence of the council's determination not to consider 'capitalization' (switching capital spending to the revenue account) as its financial position worsened was inescapable. A dwindling revenue account, with central government refusing to intervene, meant that statutory notices of redundancy would have to be activated. The council claimed that this was just a tactic to buy time for a negotiated settlement with central government, which would allow workers to be 're-employed'. As the drama unfolded, however, the split between the white- and blue-collar unions over the council's strategy – with the former generally against and the latter largely in favour – was transformed into almost total opposition when it came to the serving of the redundancy notes – by taxi, to get around the workforce's refusal to cooperate. The imagery of this exercise provided an opening for the damaging attack on 'Militant' and the City Council by the Labour Leader, Neil Kinnock, at the Labour Party Conference – characterizing the situation as one of 'grotesque chaos'. With trades union support draining away and a national party leadership firmly opposing 'Militant' in general and leading Liverpool City councillors in particular, the writing was on the wall for the politically isolated and beleaguered City Council.

The council also ran into serious difficulties on some of its outer estates. Two episodes stand out, both occurring in Speke and involving the relationship between community-based and class-based politics. Both forcefully underscore the analysis by Gyford (1985, p. 74) of the difficulties that 'community' raises, particularly 'on the Labour Left where class and community are not accorded equivalent conceptual weight'. The first concerned an initiative to unite community groups in a local community action forum. The initial inspiration came from a priest who saw the need for a festival in which residents could celebrate their local identity when, in the late 1970s, unemployment was beginning to threaten the local social fabric. The first festival ('Speke Together') – with its bands, parade, fun fair and displays – was a great success. It became an annual affair. Its resource base – provided by church and voluntary groups, the police and companies with local factories (including some multinationals) – was used to establish a broader-based community initiative, 'Speke Together Action Resource' (STAR). The aim in the words of the local priest was 'to encourage people to get into their own agenda' on such issues as unemployment, new technology and the future of work. The second festival in 1980 marked a new stage in the campaign, which shifted to developing employment opportunities through small businesses and producer cooperatives and training. Other projects included a young persons' housing cooperative, a locally based maintenance, repair and gardening service, and assistance to a credit union (to counter the operations of 'loan sharks').

The leaders of the initiative were careful not to appear to undermine the political agenda of the local Labour party. Initially, they had the approval of local politicians. However, the situation changed as the politics of the City Council shifted leftwards. The new political leadership mistrusted the involvement of local companies in STAR. Overtures were made to the organizers of the initiative to drop industry funding. These were ignored and the initiative became involved in a debate on community representation and political action. The politicians argued that the cause of the people of Speke was best advanced by the actions of their elected representatives and 'the party', and not by STAR-type initiatives. The STAR organizers argued that the crucial political significance of the initiative was that it energized local people to develop their own ideas in ways that might encourage them to adopt the agenda of the local Labour Party – 'not as passive recipients, but as actors' (Interview, 1987).

Support for the initiative had already waned with, for example, the police shifting back to 'conventional policing' in the aftermath of the 1981 riots and factory and company closures reducing industrial involvement. However, the increasing politicization of the situation speeded up this process with trade unions, local Labour party and local industry (notably Ford) sponsors all withdrawing. A major blow was the City Council decision to withdraw financial support for a 'flagship' STAR initiative – a training centre – which was then terminated by the consequent loss of EEC funding.

STAR soldiers on, but on nothing like the scale and reach of the original idea. However, the issues surrounding it remain important. They concern the question of political empowerment and the politics this demands – preceptual or a wider politics of inclusion? The founder of STAR now accepts that it might have been politically naive to involve local industry when its plant closures and redundancies were at the heart of the area's problem and were polarizing political positions. Yet he still feels that the basic rationale of the initiative – the development of wide-based community involvement – remains sound and can challenge political 'vanguardism' and the manipulative excesses into which such politics can degenerate.

The second political conflict in Speke concerned the implementation of the Urban Regeneration Strategy. Speke was one of the Programme's Priority Areas. Housing was demolished, unpopular maisonette blocks 'top-downed' and new two-storey housing built. These schemes were welcomed by the Speke community. However, plans for a sports and leisure development encompassing central Speke Park aroused disquiet and then full-blown opposition.

The plans involved the closure of the local community centre and adventure playground, key centres in Speke's community life. The community centre (or 'Commy' as it is known locally) is funded by the local education authority and had a solid base of local support – which was to prove crucial in the struggle over the local plan. This plan involved demolition of the building and the relocation of the centre's activities on separate sites. The centre workers felt this would undermine the Commy's role as a broad-based focus for community activities. The adventure playground (the 'Venny') would be integrated into the landscaped park so that the play facilities would no longer be clearly demarcated. The Venny workers felt this would endanger their carefully developed relationship with local children.

The council underestimated the depth of support for these facilities. Its position worsened when, according to a Venny worker, a leading councillor addressed a local meeting and informed it: 'this is not a proposal. This is what you're going to have!' This unsubtle exercise in preceptual politics provoked the formation of a local action group to defend the Commy and the Venny. A court injunction to prevent council access to the Commy was obtained and the building was occupied (the Venny building was owned by a charity, so its occupation was not necessary). The community 'sit-in' lasted for five months. The action group ran all the Commy's activities, even hiring blow heaters when gas and electricity were cut off. Such was the strength of local feelings that:

> Some of the Action Group never saw their families. They had their Christmas dinners here . . . We were determined to win. This place is important for the people. It's the only community centre within five or six miles. There's nowhere else for the people – where else can they go? All the groups using this place have been here since 1963, when it was opened. In some cases, five – yes, five – generations have used this place. It's home for them. (Interview, 1987).

Links were made with other groups in conflict with the council including the Black Caucus. Council meetings were disrupted and, in time-honoured Liverpool fashion, the offices of leading councillors occupied. Other council facilities in Speke – including the municipal swimming baths – were also picketed.

The Action Group was made up mainly of members of the local ward Labour Party, which divided into those supporting the council and those behind the campaign. This placed many of the Action Group in a painful and difficult situation. As already noted, the housing element of the Urban Regeneration Strategy had been very popular:

> Old houses have been upgraded, central heating has been put in. They pulled down the flats and maisonettes. There's a beautiful new semicircle of houses now, opposite the shopping centre. Have you seen all those fancy walls and wrought-iron gates? They built bungalows for pensioners. That's Hatton and company for you. They've done a terrific job in that respect. Give credit where it's due. It's sad that they felt we were dispensable. But we weren't – we just fought the buggers back. I was proud of this community the way it fought over the Commy. (Interview, 1987)

Even when the Action Group was successful and the Commy was reprieved by the Labour councillors who took over from those removed from office, the relief was tinged with a certain sadness;

> The campaign has been worthwhile. The sad thing is that, for many of us, we were fighting against our own people in the Labour Party. But this decision has proved to us that the city council is now prepared to listen to people in the community. (A member of the Action Group, quoted in the *Liverpool Echo*, 3 September 1987).

Even Knowsley, with its less preceptual style of politics, has not avoided internecine political strife involving the Unemployed Centre in Kirkby. When first established its workers were funded by the council. As central government financial restrictions started to bite, non-statutory spending necessarily came under review. Having already been funded for two years, sullied by critical press coverage and a target for local political axe-grinding, the Unemployed Centre was vulnerable. In the event, it suffered a disproportionate share of the voluntary sector funding cuts, amid much local controversy. An occupation of a council meeting by supporters of the centre added further fuel to the national Labour Party's demands for the suspension of the Knowsley North Labour Party. The national Party had instigated disciplinary procedures against a number of members. Some of the charges related to alleged membership of 'Militant' and others to 'bringing the party into disrepute' through actions relating to the imposition of a parliamentary candidate by the National Executive Committee.

The occupation of the council chamber resulted in the suspension from party membership of some leading activists. But it also encouraged a shift by activists towards direct participation in the council. One worker at the centre had already been elected to the council and the council cuts provoked another worker to successfully stand in a by-election as an Independent candidate against a prominent, long-standing councillor and former Mayor. Whether this represents the beginnings of a radicalization of Knowsley Council politics remains to be seen, but it is certainly a reflection of the localized political skirmishes that have punctuated the area's response to recent economic restructuring.

5 Conclusions

This chapter has focused on a 'locality' that has experienced a profound economic restructuring in recent years, exploring the differing responses of two local authorities to this situation. The reason for this variation lies in the contrasting political situations through which the impact of the restructuring has been mediated in the two authorities.

The responses were undercut by political opposition within the two authorities. The examples of this internal conflict presented in this chapter reinforce the theoretical point that spatial structure can actively condition political action – the spatial and social being dialectically interrelated. Thus, the spatial segregation and concentration, in the examples discussed, of ethnic and – the adjective itself captures the sense – *community* groups both facilitated politicization and helped to sustain solidarity through the close social networks that such a geography can promote.

Locality studies raise important questions of political practice. The processes of social and economic change that operate on localities are, of course, not specific to individual localities, and an understanding of this needs to inform political action. However, the impact of these processes is inscribed in the day-to-day lives of the people in those localities. Indeed it is precisely these day-to-day lives that in a sense define the most basic geography of citizenship. So any effective political response must itself be rooted in that geography. 'Building from the bottom' has to involve not just a rallying cry for the faithful but a genuine commitment to local political empowerment. This chapter has examined some of the difficult issues that such a politics raises. Thus, on Merseyside for example, the Left has been confronted with the problems of reconciling a politics of class with ones of 'community' and race. And it is reassuring to note recent signs of an increased political sensitivity to these issues. For example, the current Leader of the Liverpool City Council has publicly renounced the council's previous racial policies and is adopting a much more consensual approach generally, including that towards housing cooperatives.

At the time of writing, however, the Right is calling the political shots nationally and, as outlined in the Introduction, the overall thrust of its politics is the opposite to that argued for here and elsewhere in the book. It involves the centralization of political power and the relative disempowerment of localities. On Merseyside, this can be seen in the tight restrictions on local government spending and in the restructuring of the area's planning framework through such extra-local government initiatives as the Merseyside Development Corporation, the Speke Enterprise Zone and the Freeport.

However, not all central government intervention is operating to this formula. For example, in Knowsley, central government funding (through the Estates Action programme) has supported the local authority in a pioneering community-led environmental and housing refurbishment scheme on the Tower Hill estate. Central government funds have also been used to help set up housing cooperatives in Kirkby and elsewhere.

Yet, given the overall centralizing thrust of government policy, the motives behind these initiatives are open to question. The promotion of housing cooperatives, for example, must be viewed in the light of the government's aim to 'break-up' large council estates and reduce local government control of housing. But will such policies have the desired effect, namely the fostering of more individualized, conservative politics? It depends in part on the varying social, economic and political composition of the different localities in which the policies are tried out. For example, it is highly doubtful that government support for housing cooperatives in Kirkby will seriously undermine the area's collectivist ethos – the very ethos on which the cooperatives are themselves based. Indeed, government funding of such 'bottom-up' initiatives as the housing cooperatives and the increased involvement of local communities in planning, may reinforce collectivist politics, as in the Tower Hill scheme, by building up the confidence of local people in their ability to plan for their own locality and by encouraging the formation of community groups to institutionalize this involvement.

However, local empowerment has to involve more than simply increased participation in the political and policy-making process. It has to be backed up by the decentralization of power over resources. In Merseyside's case, the scale of resources required is considerable. A massive programme of social and economic reconstruction is needed and this will require substantial public funds. An indication of the required scale is provided by the proposed 'Merseyside Integrated Development Operation', a public-sector-led scheme for a five-year reconstruction programme involving a major upgrading of the environment, infrastructure and communications, support to local industry and services, and a major training programme. The programme is costed at £330 million. It is estimated that around 23 000 jobs will be created, but even this represents only about one-sixth of the number of jobs that were lost in the Liverpool 'Travel-to-Work Area' in the 1970s and early 1980s.

So the initiative represents only a first, albeit important step in the social and economic regeneration of the area. However, it is an important testing-ground for the politics of urban regeneration. The public sector on Merseyside is represented in the initiative by all five district councils (see Figure 4.1) and the North West Water Authority. The Merseyside Development Corporation is involved, and central government is participating through its Merseyside Task Force. The plan is to attract about one-half of the funding from the European Social Fund and the European Regional Development Fund and the other half from national public and private sources. As public in this context has to mean mainly from central government, the seriousness of the government's stated commitment to urban regeneration will be tested – both in terms of the amount of money provided and the discretion allowed to local administration and spending. But there is also a more important longer-term political issue where both central and local government will be tested. This concerns the extent and nature of local community involvement in the programme. Will it build from the bottom or impose from the top?

Acknowledgements

The work was carried out while the author was at CES Ltd, in collaboration with Jane Lewis (formally of the Department of

Geography, Reading University, now with the Economic Development Unit of the London Borough of Ealing). Research assistance was provided by Frank Banton (Ockenden Venture, Liverpool), Bernard Charles SJ (Liverpool), Heather Clark and Ann Smith (Training for Development, Liverpool), Maggie Pearson (Department of General Practice, Liverpool University), Brian Bailey and Ian Thompson (Department of Civic Design, Liverpool University) and Lorraine Donnelly (Liverpool Polytechnic). The usual disclaimers absolving these individuals from responsibility for any errors of fact and interpretation apply. Finally, thanks are due to all the people we interviewed, whose patience we tried but seemingly never exhausted.

References: Chapter 4

Andrews, G. 1985. 'The greening of Liverpool.' *Guardian*, 2 December 1985.

Architects Journal 1988. 'The light at the top of the tunnel: community-led regeneration by the Eldonians,' 23 March, 37–63.

Ben-Tovim, G. 1989. 'Race, politics and urban regeneration: lessons from Liverpool.' In M. Parkinson, B. Foley and D. R. Judd (eds) *Regenerating the cities: the UK crisis and the US experience*. Glenview, Ill., Boston, Mass., London: Scott, Foresman and Company.

CDS. 1987. *Building democracy: housing cooperatives on Merseyside*. Liverpool: Cooperative Development Services.

CES Ltd. 1988. 'People and places: a classification of urban areas and residential neighbourhoods.' *CES Paper 33*, London: CES Ltd.

Crick, M. 1986. *The march of Militant*. London: Faber and Faber.

Daniels, P. 1986. 'Producer services and the UK space economy'. In R. Martin and R. Rowthorn (eds) *The geography of de-industrialisation*. London: Macmillan.

Grosskurth, A. 1985. 'Bringing back the Braddocks.' *Roof*, January/February, 19–23.

Gyford, J. 1985. *The politics of local socialism*. London: Allen and Unwin.

Kilroy-Silk, R. 1986. *Hard Labour*. London: Chatto and Windus.

Lane, T. 1987. *Liverpool, gateway of Empire*. London: Lawrence & Wishart.

Liverpool Black Caucus. 1986. *The racial politics of Militant in Liverpool: The black community's struggle for participation in local politics 1980–1986*. Liverpool: Merseyside Area Profile Group and Runnymede Trust.

Liverpool City Council. 1983. *Liverpool City News: a special Council Bulletin*, November 1983. Liverpool: LCC.

Liverpool City Council. 1984. *Liverpool population profile*, City Planning Officer, April 1984. Liverpool: LCC.

Liverpool City Council. 1986. *The strategic approach: urban programme, 1986–87*. Liverpool: LCC.

Lloyd, P. E. 1970. 'The impact of development area policies on Merseyside, 1949–1967'. In R. Lawton and C. M. Cunningham (eds) *Merseyside: social and economic studies*. London: Longman.

Macdonald, A. 1986. *The Weller way: The story of the Weller Street Housing Cooperative*. London: Faber and Faber.
Meegan, R. A. 1988a. 'Unfulfilled promises: the growth and decline of Merseyside's outer estates.' In P. N. Cooke (ed.) *Localities*. London: Unwin Hyman.
Meegan, R. A. 1988b. 'Economic restructuring, labour market breakdown and locality response.' In J. Morris, A. Thompson and A. Davies (eds) *Labour market responses to industrial restructuring and technological change*. Brighton: Wheatsheaf Books.
Merseyside Council for Voluntary Services. 1978. 'One in eight: a report on eleven outlying housing estates where one in eight of Merseyside's people live', Liverpool: Merseyside Council for Voluntary Services.
Office of Population Censuses and Surveys. 1988. *OPCS Monitor*, PP3 88/1. London: OPCS.
Smith, D. M. 1969. *Industrial Britain: the North West*. Newton Abbot: David & Charles.
Taafe, P. and Mulhearn, T. 1988. *Liverpool: a city that dared to fight*. London: Fortress Books.
Urry, J. 1981. 'Localities, regions and social class.' *International Journal of Urban and Regional Research*, **5**, 455–73.

5

Coping with restructuring: the case of South-West Birmingham

DENNIS SMITH

1 Characteristics of the locality

In order to understand the response of South-West Birmingham to the challenge of political and economic restructuring it is necessary to know something about the peculiarities of this locality. South-West Birmingham consists of the eight contiguous wards of Selly Oak, Bournville, Weoley, Bartley Green, Brandwood, Kings Norton, Northfield and Longbridge and has a population of approximately 200 000. These suburbs of Birmingham stretch out on either side of the Bristol Road, which dissects the locality diagonally from Selly Oak to Longbridge.

Major landmarks are the Bournville works of Cadbury Schweppes, to the east of Bristol Road just south of Selly Oak, and the Longbridge plant owned by Austin Rover. The 'Austin' is in the south-west extremity of the locality close to the point where the Bristol Road takes a sharp right turn into Rubery. South-West Birmingham has clear boundaries to the north, south and west, provided by countryside and the campus of Birmingham University. To the east the boundary is more arbitrary.

The locality has a distinctive housing pattern. In the city of Birmingham 53 per cent of housing is owner-occupied (OPCS, 1983). All but three of South-West Birmingham's wards register lower percentages. However, the figures for the wards of Bournville, Northfield and Selly Oak are 57 per cent, 60 per cent and 68 per cent respectively. In the city as a whole, 35 per cent of dwellings are rented from the council. The three wards just mentioned do not reach this level. In Selly Oak council property constitutes a mere 12 per cent of the total. However, in marked contrast, Brandwood (42 per cent), Longbridge (47 per cent), Weoley (51 per cent), Kings Norton (60 per cent) and Bartley Green (63 per cent) all contain levels of council housing greatly above the city average.

In five of the wards the proportion of employed people in the semi-

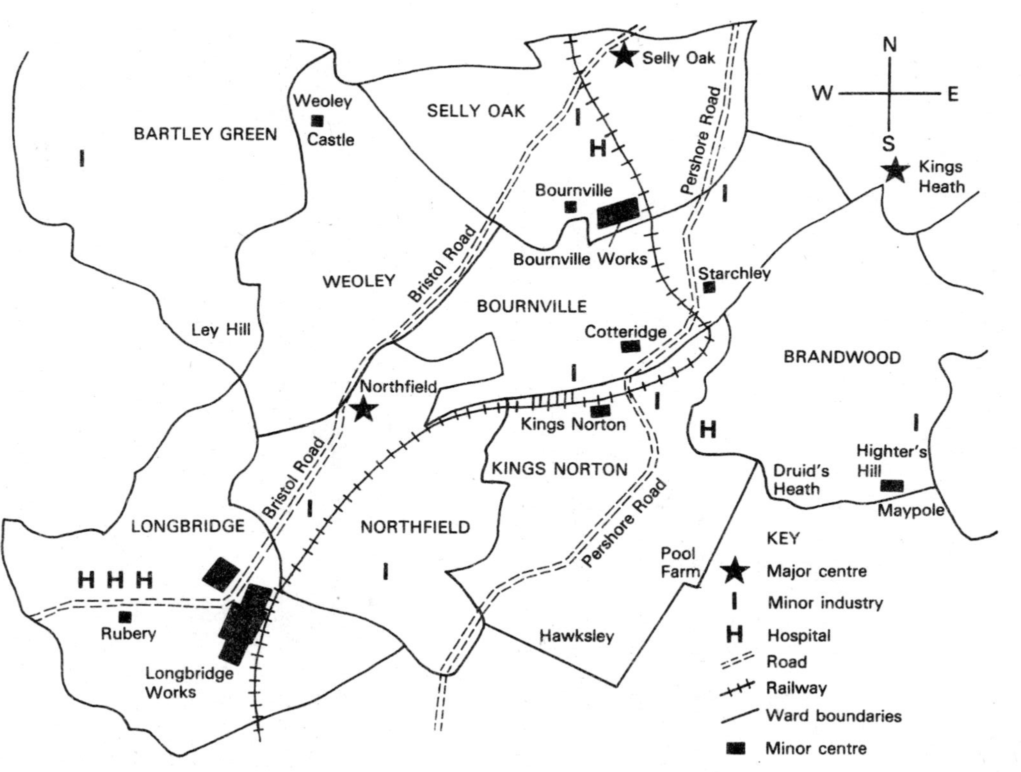

Figure 5.1 South-West Birmingham.

skilled category is higher than the figure of 22 per cent for the whole city. In Longbridge it is as high as 28 per cent. Only one of the wards (Longbridge) has a smaller proportion of non-manual workers than the city-wide rate of 26 per cent. In Selly Oak, Bournville, Northfield and Brandwood the proportion is 30 per cent or slightly above. Skilled and self-employed workers varied by ward between 21 per cent (Bournville) and 26 per cent (Longbridge) of the employed population: the city-wide figure is 24 per cent (1981 Census).

Austin Rover and Cadbury Schweppes provide approximately 13 500 jobs between them. A substantial proportion of these jobs are held by residents of South-West Birmingham. Just over 84 000 of these residents were in the labour market in 1981. In 1986 a postcode analysis showed that more than 40 per cent of Austin Rover (primarily male) hourly-paid employees lived in postcode districts falling mainly in South-West Birmingham. The same applied to Cadbury shopfloor workers. Approximately 80 per cent of women workers lived in the locality, as did about 65 per cent of the males. More detailed analysis revealed that the heaviest concentrations of Cadbury and Austin workers living locally were to be found in the residential developments extending north and south from the Bristol Road throughout its length.

Like the rest of the city, South-West Birmingham has shared in the economic downturn that has hit the West Midland region since the early 1970s. The impact in South-West Birmingham has been felt on local industry and local public services, including health and housing. The locality has not organized as a 'constituency' in order to mount a defensive strategy. Instead, there have been a few highly localized responses designed to help families cope with the problems of the recession. These have been slow to develop. Similarly, acknowledgement at the higher levels – the city and the region – that the 'affluent' and individualistic West Midlands would have to organize to defend its interests was belated. Before developing this argument further, some aspects of the historical background will be outlined.

2 The historical legacy

During the late nineteenth century local government in Birmingham acquired a reputation for providing a wide range of social facilities, such as libraries, schools and swimming baths, for its citizens. The city's boast was that under the leadership of a civic-minded professional and business class economic efficiency and social welfare were effectively combined. Joseph Chamberlain was a powerful propagandist for this view. In fact, Birmingham's 'civic gospel' came unstuck on at least two points. The costs were higher than anticipated – and this upset the solid businessmen on the aldermanic benches. And the benefits to Birmingham's ordinary working families, especially in the sphere of housing, were much less than were promised.

In some respects at least, the factory and village established by Richard and George Cadbury during the late nineteenth and early twentieth centuries at Bournville, a few miles south west of the city centre, were a well-advertised reproach to the Chamberlainite faction and a demonstration that business sense and social welfare could indeed be effectively combined – at the level of the neighbourhood rather than the metropolis. The Cadbury worker was to be self-disciplined, healthy, punctual, obedient and – if he or she had the capacity – enlightened (Dellheim, 1987).

During the twentieth century, the Bournville ideal of enlightened paternalism leading to responsible citizenship has been overtaken, on the one hand, by the growth of the national welfare state and, on the other, by the shrugging off by business of any particular active concern with nurturing the local community. These processes are expressed geographically in the growth of Birmingham to the south-west of Bournville – a vast area dominated by council estates and the motor works at Longbridge.

A century and a half ago the open fields and canals of the locality were all encompassed within the ancient parishes of Kings Norton and Northfield. The Birmingham and South-West Suburban Railway, opened in 1876, helped to bring new labour and capital into the locality. Passengers from the centre of town were soon alighting at Selly Oak and Bournville. From 1879 the Bournville site was occupied by the new factory owned by the Cadbury brothers. The Bournville works dominated the north-eastern part of the locality. From 1905 its south-western part was under the influence of the motor company established at Longbridge in that year by Herbert Austin.

Austin did not try to emulate the programme of all-round community improvement being promoted in Bournville, where George Cadbury had established a model village with parks, recreation grounds, sports facilities, libraries and a wide range of educational institutions. The car maker was much more interested in the levels of physical effort and technical skill that the worker brought into the business than the extent to which capital could use its power to raise the moral tone of labour (Church, 1979).

From the 1890s onward, South-West Birmingham drew in both capital and labour, both in search of a better deal than they could get in the crowded inner city areas of the Midland towns or the depressed areas of northern England, Wales, Scotland and Ireland. Skilled artisans bought terraced houses in Cotteridge, Stirchley, Bournville and Selly

Oak – and further south in Northfield and Kings Norton, both of which had been included in the city by the boundary extension of 1911. The expansion of the car industry between the wars encouraged rapid residential development.

When Roland Cartland was adopted as its Conservative parliamentary candidate late in 1933, Kings Norton included both 'the Austin' and the Cadbury works at Bournville. The following year, he described the constituency as follows:

> The Division consists of three wards, Selly Oak, Kings Norton and Northfield. The Northfield Ward, the largest in area in Birmingham, runs the entire length – north to south – of the Division. Until a few years ago it was entirely countrified with only a number of villages and farms making up the electorate. The Austin Works situated at the extreme end of the Ward and a small housing estate near to – that was all. Today an almost unbelievable development has taken place. Two vast Corporation housing estates (of over 2000 houses each) have been built in the Ward, and private enterprise has built and is building not only along the main roads, but in the villages and in what seemed almost inaccessible spots. The character of the Ward has completely changed. (Cartland, 1942, p. 93).

Even more huge council house estates, some with high-rise flats or blocks of maisonettes, have been built to the west and south in the locality since the last war (Sutcliffe and Smith, 1974). Interwoven with council property are clusters of private development, for example in Bartley Green to the west. Some of these private dwellings are very handy for commuters using the M5, M6, M40 and M42 motorways, which skirt South-West Birmingham.

The local culture of South-West Birmingham expresses in intensified form sentiments to be found throughout the city. It is a culture in which people believe that what really counts is getting a nice home of your own and having enough money to enjoy it. It is commonly assumed that although you can trust a few people you know very well you should be very careful of everybody else. Politicians are held to be shifty individuals who are only after your vote. They do not, it is thought, live up to your expectations – if you had any, that is. The watchword is: look after yourself because you're the only one who will.

This culture is as much a historical product as is the industrial structure. Improvement or 'getting on' has been the main objective of the people who have come to work in the area during the last century. Despite the paternalistic ethos embedded in Bournville's past, other routes to improvement have been more popular. South-West Birmingham as a whole was built upon the principle of getting on by looking after yourself. It is not really tuned in to ways of combining getting on with looking after one another. Local people have chosen what seem to be the most effective ways of surviving in the world as they find it. They work well within small groups where everyone puts something in and everyone gets something out. This is the heartland of the darts league and the angling club. This spirit is conveyed in the following reminiscence of a skilled tradesman who joined the Austin in 1946:

> I found out I was the first outsider that they'd employed [in that particular section] for years. They were usually relatives – fathers, uncles, cousins. [The] chargehand was testing me out because he didn't want an outsider. He did his best for me to fail but fortunately I succeeded . . . After I'd put this [particular machine] together he said 'You'd better come . . . in the pen now'. I'd served my apprenticeship, so to speak. He told me which bench to work on . . . I found out —— was the ringleader of that side of the shop. [He] said 'You're on the best side of the shop. You fit in with us and you'll be all right. Have nothing to do with that other lot.'

If it is to be effective, policy has to be sensitive to the modulations of local culture. It is relevant that local people tend to be cynical about the big battalions – the town hall, the large employers – even though they accept that they cannot resist their power. The inhabitants of South-West Birmingham, as elsewhere in the city, rather resent the idea that you have to 'sell' yourself or your locality to strangers, especially those with capital to invest. In fact, until the 1960s, South-West Birmingham was a place that everybody else (or so it seemed) was keen to move into as part of the universal quest for a better job and a better life. Keeping out people you do not want has preoccupied locals more than getting in people you want.

There is, it should be added, a slight undercurrent of testiness, verging on intolerance, with respect to Afro-Caribbean and Asian settlement in Birmingham, although the south-west segment of the city remains overwhelmingly white. This sentiment was politely expressed as follows by one woman resident of Bournville: 'I shouldn't be naughty, but I think our coloured friends have taken over a bit . . . I'm not prejudiced because I know you've got to work with them, but . . .' In fact, Asian and Afro-Caribbean employees are to be found at both Austin Rover and Cadbury Schweppes, for example on the night shift at the latter factory. Relatively few of these employees live in South-West Birmingham.

3 The slump

Since the early 1970s, South-West Birmingham has shared in the shift from boom to slump in the West Midlands. Between 1969 and 1981 the numbers employed in manufacturing in the West Midlands county fell from 837 774 (based on 1958 SICs) to 499 858 (1968 SICs). The rise in service-sector employment – from 452 914 to 623 514 (Department of Employment, Annual Census of Employment) – did not compensate for this decline. In the early 1960s regional gross domestic product (GDP) per head of population was nearly 10 per cent higher than for the entire United Kingdom. Only the South-East was performing more successfully. By 1976, the West Midlands had fallen back to the level of Humberside, Yorkshire, the East Midlands, the North of England and Scotland (Liggins, 1977). Between 1976 and 1981, GDP per head in the West Midlands declined more rapidly compared with the UK average than any other region. By 1981, the West Midlands was the poorest region in England (Spencer *et al.*, 1986, p. 64). Two years later, the government created an unofficial 'Minister for the West Midlands'.

In October 1987, more than 35 per cent of all Birmingham's unemployed had been out of work for more than two years and more than 51 per cent for more than a year. For men only, the comparable figures were more than 39 per cent and more than 55 per cent (City of Birmingham Development Department, 1987). During the twenty years up to 1969, less than one economically active person in a hundred was out of work in Birmingham. By 1981, unemployment had risen to above 10 per cent in all the wards of South-West Birmingham. By 1986, about one in five of all potential workers was out of a job. During 1987, the level of unemployment fell to about one in six in the worst-hit wards of Longbridge, Kings Norton, Bartley Green and Weoley. Selly Oak and Northfield were doing rather better (one in eight), but not so well as the national average of about one in ten.

In the face of this situation, the West Midlands, 'for long unconscious of its own territoriality' (Morgan, 1985, p. 566), has been slow to get its act together and demand assisted-area status (West Midlands County Council, 1981; West Midlands TUC, 1982). Attempts to establish an effective 'spatial coalition' of business, unions and councils were weakened by local divisions, although there have been some moves in this direction: for example, some informal help by individuals promoting Birmingham business abroad (and accredited by the City Council) to the local car industry in its search for export markets in the Far East. However, in 1982, the *Financial Times* wrote that the West Midlands had one of the 'most ineffectual lobbies in the UK' (2 November 1982). The regional CBI and the Birmingham Chamber of Industry and Commerce were both opposed to the Labour-controlled County Council's project of a regional developmental authority, which would be able, like the present West Midlands Enterprise Board, to buy equity. As Morgan (1985) comments, 'the principal "actors" in the West Midlands were resolutely partisan with respect to remedial action' (p. 568). Although the abolition of the West Midlands County Council has removed one of these actors, divisions remain.

One strategy adopted in response to the slump has been the deliberate courting of inward investment from overseas. During 1986, the area covered by the West Midlands Industrial Development Association attracted 80 out of a total of 230 new investment projects in Britain by foreign companies (*Choice*, 15 April 1987). Telford, for example, was recently dubbed the Tokyo of the West Midlands in the local press.

Moves have also been made in the direction of increasing local exports. In October 1986, Birmingham City Council produced a directory entitled *Made in Birmingham*. Of the 30 000 copies produced, 80 per cent were directed to actual or potential overseas customers (*Birmingham Post*, 24 October 1986). In a report at the end of 1986, Birmingham City Council's Development Department indicated a wide range of local successes achieved during that year: investments by locally based companies such as Cadbury Schweppes, IMI and Foseco Minsep; city-centre office and shop developments; inner-city programmes, such as the Aston University Science Park and the Innovation and Development Centre; the establishment of the Woodgate Business Park (in South-West Birmingham) as part of the council's industrial land strategy. Apart from the £120 million spent on Birmingham's International Convention Centre, some £30 million was spent by the council on city economic projects during the year.

A wide range of funding sources was drawn upon, including the Inner City Partnership Programme, the European Social Fund, the European Regional Development Fund and the Manpower Services Commission. More than £60 million was levered from the private sector (*Birmingham Post*, 2 December 1986).

Within the Development Department, the Economic Development Unit is currently offering local or incoming businesses advice about a range of services, including a new technology training scheme, consultancy at the Birmingham Microsystems Centre, the city's wage subsidy scheme, marketing support, a number of new enterprise workshops, and sites within economic regeneration areas under the terms of the city's economic and industrial land strategies. The 1986 review of the economic strategy included plans for a Composite Materials Centre, an Automotive Engineering Centre and a Biotechnology Centre. The last three initiatives are of obvious relevance to South-West Birmingham's major employers, Austin Rover and Cadbury Schweppes.

The most significant local public venture in South-West Birmingham has been the opening of the 60-acre Woodgate Business Park in the north-western part of the locality during 1986. However, this represents a relatively small level of investment compared with the planned activities of the new urban development corporation established early in 1988 for East Birmingham, an area of more than 2000 acres mainly in Aston, Nechells and Washwood Heath.

4 Restructuring South-West Birmingham

Despite the tardiness and lack of coordination in the West Midlands' regional response to the slump, and despite the fact that South-West Birmingham has not been made the object of special development programmes (unlike East Birmingham), the impact of change in the locality has been less than might have been expected.

Restructuring has been relatively unopposed for two main reasons: first, the sympathetic response by many local people to the strong political appeal to individualistic/ratepayer/home-owner interests, which has accompanied restructuring over the past decade and a half; secondly, the absence of organized and articulate expressions of solidarity based upon class or locality.

In South-West Birmingham the pain of change has been alleviated in three ways. First, by the relative success of the locality in attracting new service-sector employment to replace lost manufacturing jobs. Specifically, between 1971 and 1984 more than 118 000 jobs in metal engineering were lost in the Birmingham travel-to-work-area (TTWA) – representing a reduction of the total employment in this sector by 44 per cent. Only about 26 000 service-sector jobs were gained – an increase of 7 per cent. By contrast, in the travel-to-work-area that includes South-West Birmingham (that is, the TTWA covered by employment offices in Selly Oak, Northfield and King's Heath), nearly 16 000 metal and engineering jobs were lost (a reduction of 42 per cent) but more than 11 000 service-sector jobs were gained (an increase of 30 per cent).

A second way in which the penalties of change in the locality have been reduced is by the relatively high levels of investment in new technology made by Austin Rover and Cadbury Schweppes. At Longbridge the £100 million investment in the Metro line, beginning in the late 1970s, was followed by further projects in collaboration with the Japanese company, Honda, culminating in a decision to go ahead with the design and manufacture of a new family car (the R8). At Bournville, new machinery for production of the Wispa chocolate bar was at the centre of a programme costing at least £100 million during the 1980s. A more recent example of the company strategy of upgrading its capital equipment was the commissioning of a new creme egg plant in 1987. At the same time, it may be added, the company has built up a reservoir of temporary labour (constituting very roughly 10 per cent of the work force), which it can employ flexibly and at relatively low cost.

It is difficult to be sure why South-West Birmingham was favoured for new investment by the two companies, rather than competing sites in other parts of the country. In the case of Cadbury Schweppes, the very strong local ties of the Cadbury family may have been relevant. In the case of Austin Rover, a company owned at the time by the government as major shareholder, the marginal electoral seat of Northfield could have been a consideration.

A third factor has been the increasing importance of Longbridge and Bournville within the national operations of Austin Rover and Cadbury Schweppes. This must be put in context. The number of hourly-paid employees at Longbridge fell from 19 000 in 1977 to just over 11 000 in 1982. Cadbury's shopfloor workforce declined from about 7000 in 1980 to about half that amount by the mid-1980s. In conjunction with developments elsewhere in the two companies, these changes meant that by 1984 each of the firms drew more than one-third of its national workforce from South-West Birmingham. While the companies remain in business in their present form, they seem likely to be heavily dependent upon the South-West Birmingham labour force.

The point is partly psychological. Local employees perceive that local operations constitute a high proportion of the total undertaken by the company in each case, and find it difficult to acknowledge that the Austin or the Bournville works could close down. The recent decision to shut down one of Austin Rover's Cowley factories at Oxford, following the takeover by British Aerospace, is likely to have strengthened this perception, in the short term at least. All such perceptions and expectations are, of course, liable to prove false.

Despite the factors mentioned, economic restructuring and redundancy have been associated with a great deal of painful material and psychological adjustment in the locality. Much of the suffering has been rendered invisible. The unemployed disappear from view, socially and politically. Employees still in work have been glad to have a job and generally ready to acquiesce in tougher regimes on the shop floor. Wage settlements have also tended to edge upwards during the early phase of the third term of the Thatcher government.

However, the isolation of the unemployed and the frustration of employees faced with greater pressures to toe the line (acceptable to many as part of the overall 'job bargain') are only two of the penalties imposed by restructuring. Tensions are being generated on a much wider scale, both within individuals and between groups.

In the rest of this chapter attention will be paid to policy responses to some of these tensions in the sphere of social reproduction, especially health and housing. The chapter will end with a brief survey of ideology and practice in a number of contrasting 'coping strategies' directed at helping beleaguered individuals, families or communities. These include the local neighbourhood offices, the South Birmingham Family Service Unit and the Bournville Village Trust.

5 Social reproduction: winners and losers

The first topic to be explored is the *local health service*. Since 1971, the population in the area covered by the South Birmingham Health Authority (i.e. the wards of South-West Birmingham plus Billesley and Moseley to the east and north-east) has tended to decrease; the projected total in 2001 is nearly 27 000 fewer than at the earlier date. The population is ageing, especially in Selly Oak and adjacent Bournville. Within the West Midlands, only Hereford has more pensioners as a proportion of the total population than does South Birmingham. This is in spite of the fact that men in South Birmingham have a consistently lower life-expectancy than the regional average from birth to age seventy.

The 1981 Census also shows that South Birmingham has a proportion of single-parent households that, at 6.7 per cent, is well above the regional average. Many families with three or more children are to be found in Kings Norton and Longbridge. The relative poverty of several such families is shown by the example of Longbridge ward. More than 42 per cent of households with children do not own a car – and in the case of those households with three or more children this goes up to more than 50 per cent. More generally, the parts of South Birmingham that score highest (or 'worst') in terms of key social indicators, such as the proportions of children aged 0 to 4 years, single parent households, households lacking a car, and youth unemployment, are to be found clustered on the more run-down council estates on the fringes of the locality: in the southern reaches of Kings Norton, to the south-west and north-west of Longbridge ward, and on the boundary between Bartley Green and Weoley wards (South Birmingham Health Authority, 1986).

The South Birmingham Health Authority (SBHA) employs nearly 5000 people (excluding senior staff) and runs eight hospitals. Plans to increase maternity provision at the site on which Selly Oak Hospital is located and to transfer the Birmingham Accident Hospital there were approved by central government late in 1986. In March 1987, these developments were delayed owing to limitations upon the cash available (*Birmingham Post*, 22 April 1987). More generally, SBHA has suffered from a squeeze upon its resources. Between 1982/83 and 1984/85 its purchasing power decreased by about 6 per cent owing to the underfunding of national pay awards, higher than average price increases in relevant goods and services, and the increasing proportion of elderly people with greater health care needs (*Health Journal*, December 1986).

The health authority has conducted a sustained and powerful campaign against this cash shortage. For example, the chairman of SBHA, a prominent Labour councillor, declared in October 1986 that he had attended a seminar on health care that showed 'that Birmingham is the most deprived part of the country – and that parts of Birmingham are among the most deprived in Europe'. A consultant on the South Birmingham district medical committee stated that all Birmingham's major hospitals would have to close two or three wards in 1987 if the financial problems were not solved (*Birmingham Post*, 30 October 1986). Three months later, the SBHA's general manager accepted that, as the local paper put it, 'There are more patients in South Birmingham waiting longer to be admitted to hospital than anywhere else in England' (*Birmingham Post*, 24 January 1987). This comment applied especially to people waiting for ear, nose and throat treatment and for orthopaedic operations. In all, more than 58 per cent of patients had been awaiting treatment for more than a year.

Health policy as it affects South-West Birmingham veers between the magnetic pulls of, on the one hand, the 'fundamental review' under way and, on the other, the intense pressures being exerted for short-term cash injections. In May 1987, SBHA appointed a new district medical officer. Dr David Arnott Player, the appointee, had been director general of the Health Education Council, whose report *The Health Divide* unfavourably contrasted health provision and outcomes for the poor as compared with the professional classes in Britain. The Health Education Council was, shortly after publication of the report, brought under government control as the Health Education Authority. As the foregoing facts imply, the orientation of SBHA remains oppositional.

Another major arena of competition for resources is *housing*. The link to health and welfare issues was explicitly pointed out in a report of the Housing Committee to the Birmingham City Council. The committee stressed the great impact upon its work of restrictions upon spending, including the moratorium placed by the Secretary of State (in October 1980) on new building contracts and council grants and loans for improvement and purchase. The report argued that:

> the major impact will be felt . . . by some of the most vulnerable and disadvantaged groups in society who are increasingly to be found in Council accommodation or in need of it.

> Council and housing association tenants are comprised of a far greater proportion of families with special needs than any other sector of the housing market. Firstly, those becoming Council tenants as a result of homelessness or as priority from the Waiting List are those who generally through low incomes are least able to compete in the declining private rented sector or in home ownership. Secondly, those who are leaving tend to be the more stable families with higher incomes who are able to purchase their own accommodation. The combined effect of these factors is to produce a public sector comprising high proportions of families from low socio-economic groups, one parent families, large families, significant proportions of elderly and more dependent vulnerable groups who require higher levels of support and assistance. (Birmingham City Council, 1981, p. 21).

Specific problems included the fact that new building programmes had mainly been for the poor and elderly and many grants or loans had been for modernizing dwellings. The moratorium meant that old people would 'continue . . . to under-occupy family houses with out-of-date facilities and this will restrict the opportunity for larger families to be more appropriately accommodated'. Meanwhile, many young couples waiting for accommodation would move outside Birmingham, leaving the aged still to be provided for. More generally, lack of capital expenditure on the more run-down council estates would disproportionately affect the most vulnerable groups living there. Finally, the report continues,

> it is at the individual level that the full effects of the reductions in expenditure must be judged. As the Waiting List grows and the Urban Renewal Programme slows down, the longer families remain in inadequate, sub-standard or overcrowded housing conditions, the greater is the personal and family stress experienced. There is considerable evidence to demonstrate the strong associations and in some cases direct causal relationships between housing stress and social deprivation. Low educational achievement, unsatisfactory child development, family breakdown, domestic violence, child abuse, various crimes and vandalism have all been attributed in whole or in part to the quality of housing occupied (Birmingham City Council, 1981, p. 7).

A further legislative constraint upon local government in the housing field was the 'right to buy' provision included in the Housing Act of 1980, which gave most council tenants with three years' tenancy the legal right to purchase their dwellings, with the benefit of full mortgage facilities and discounts of between 33 and 50 per cent.

In combination, the moratorium on capital spending and the right to buy provisions have increased deprivation at the bottom of the housing ladder while making it easier for some to climb further up it.

In other words, these measures have increased polarization locally. This has to be seen in a broader historical and social context. Since the war, many manual workers' families (including car workers) have moved out of the worst council housing, escaping in particular from less desirable estates such as Ley Hill and Pool Farm. As one local councillor put it:

> I think the reason why people in Ley Hill or Pool Farm don't work at the Austin [nowadays is] because [these particular workers] moved out to other areas and the people who have moved in are generally the one-parent families or unemployed people who have no choice where they move . . . You would find that a lot of people who in the 1960s worked at Longbridge actually bought their houses and . . . a lot of them lived in the small private estates . . . because in those days they were earning enough to enable them to buy. So there was a tendency for them to be council tenants for a time and then to buy a house later on. I mean that's certainly my experience, when I grew up in those estates in the 1960s. (Interview, 1987)

Two other factors are relevant. The restructuring of the car industry and components suppliers has created a climate of insecurity and, in many cases, economic hardship among local residents. Mortgage repayments have been harder to meet. At the same time, one consequence of slum clearance in the inner city has been the arrival of a new kind of tenant in the council estates of the outer suburbs.

By and large, families prefer to avoid or get out of the deteriorating council housing on the southern and south-western fringes of the locality. For such families, Frankley estate (built during the late 1970s and early 1980s on the western border) would be a step 'up'. Even 'better' would be a house on one of the 'nicer' council or private estates in some parts of Bartley Green or Weoley Castle. People struggle to make the unsatisfactory bearable, as the unemployed man in his early thirties quoted below has done:

> The house was an absolute wreck when we moved in. It took over three years to do some reasonably cheap decorating . . . We have got a paraffin heater for upstairs . . . [but it] makes everything filthy. But we've got a garden which I didn't have before, and I love the garden . . . [It's] quite pleasant [here] . . . [There's been] a fair bit of trouble with local youths throwing bottles and cans over the fence . . . [but] I don't regret moving here [to Weoley] at all. The Weoley Castle people on 'Crossroads' are . . . well, you watch it. They came out with the comment that all the Weoley Castle people are rogues or unemployed. I thought: 'Oh, very nice!' . . . [I wouldn't want to live in] Northfield, it's too hemmed in. You often hear of trouble; always somebody being mugged or something in Northfield . . . I wouldn't like parts of Weoley Castle. Rubery I don't like at all. We were offered a house there. [It was] like a dungeon. (Interview, 1986)

Rubery, a very run-down area of decaying maisonettes behind the Longbridge car factory, is frequently referred to by local people in terms that suggest that they would not like to live there. At the other extreme, along some roads in Bournville and Northfield a certain quiet sense of being 'a cut above the rest' sometimes comes across. Selly Oak, close to Birmingham University, provides haunts for the more bohemian intelligentsia who, to some extent, look outside South-West Birmingham towards Moseley, Kings Heath and Edgbaston.

On a more speculative level, it is possible that the processes of restructuring in industry and the public sector have combined to increase the social tension surrounding the housing ladder, breeding a degree of resentment and community between the relatively better-off and the relatively worse-off (cf. Rex and Moore, 1967). Some of the more dramatic manifestations are suggested in the following newspaper report:

> Local police [in Selly Oak] have repeatedly asked for public help in curbing a new type of crime called 'steaming' in which gangs have used sheer strength of numbers to intimidate local shopkeepers into handing over cash and goods. [A police spokesperson] said 'I do not know why Selly Oak is a centre of attraction for youngsters from places like Northfield and Rubery but it is and I would like to know why myself.' (*Choice*, 27 November 1987)

6 Coping strategies at the local level

There is no spatial coalition looking after South-West Birmingham. Instead, there is a series of local coping mechanisms. Some of these are local branches of national organizations. For example, the South Birmingham Health Authority provides a wide range of services for coping with disturbances to individuals and families, including the emotional and psychological disruption associated with economic restructuring and redundancy. The local offices of the Manpower Services Commission also offer a number of facilities through their job centres, including community programmes, adult preparation training and restart courses. Rather than describe these services in detail I shall focus on three highly distinctive but local responses to the changing circumstances of individuals and communities in South-West Birmingham, each embodying a particular strategy of adjustment.

The responses to be discussed are those of the Bournville Village Trust, the local neighbourhood offices of the city council, and a voluntary community and social work centre – the South Birmingham Family Service Unit (FSU) based in Kings Norton. The nature of the management differs considerably between the three schemes:

(a) Bournville Village Trust is a charitable body responsible for about 7500 dwellings on more than 1000 acres with a population of about 23 000 (Hensloe, 1984). It is dominated by its senior employees – including a general manager, deputy general manager, chief accountant, housing manager, chief estates officer, and community and information officer – who manage the Trust's affairs.
(b) The neighbourhood offices have been set up throughout the city by the city council and are staffed by local government officers representing the chief executive's housing, social services and environmental health departments in a 'one-stop shop'.
(c) The South Birmingham FSU depends upon a mixture of voluntary contributions and grants from government and other sources. It is managed by an executive committee staffed equally by professional social/community workers and local volunteers. The FSU has a strong connection with residents on the Pool Farm, Primrose Hill and Hawksley estates.

By far the oldest of these agencies is the Bournville Village Trust, which was established in 1900 to foster the development of a model village and cater for the physical, educational and moral needs of a socially mixed community. The trust has passed through three phases since the beginning of the century. During the first phase, up to the beginning of the Second World War, Bournville provided a local (and to some extent national) benchmark for standards in housing and community facilities. During the second phase, between the 1940s and the 1960s, national developments in the public provision of health, education and welfare facilities (accompanied by the spread of television and car ownership) produced a dual effect in Bournville. On the one hand, the old paternalistic relationship between benevolent private capital and a specially privileged local tenant population was deeply eroded. On the other hand, Bournville residents enjoyed the leafy seclusion of their nice houses and quiet avenues. A certain cosy parochialism developed.

Restructuring disrupted this in two ways during the third phase. First, the Housing Act of 1980 not only encouraged council house tenants to buy their own homes, a possibility denied to Trust tenants, but it also strengthened the rights of tenants in general against landlords. During the 1980s, tenants of the Trust have felt aggrieved that, compared to owner occupiers, they are being asked to bear a very large portion of the cost of maintaining the estate through their rents. Several tenants' associations have come into existence. A second consequence of the pressures of restructuring has been the determination of Cadbury Schweppes, which has no legal or formal connection with the Trust on its doorstep, to sell off a large part of the land at Rowheath previously devoted to sporting and leisure activities. This area provided a green wedge in conformity with accepted planning principles. It served as a shield for Bournville residents against the noisy outside world. Many of them fought vigorously to preserve this pleasant environment and they have succeeded in restricting the extent of new building.

It is interesting to compare recent developments in Bournville with the progress of the South Birmingham Family Service Unit. During the mid-1980s, the FSU strengthened its links with the local residents, partly through campaigns for facilities such as a community nursery and a community launderette. In contrast to the Trust, which experienced the growth of tenants' associations during the 1980s as a source of unwelcome pressure from below, the social and community workers in the FSU have gone out of their way to encourage community participation. Street meetings are a common form of consultation with local residents, who are mainly council tenants. In the words of one of the latter, an unemployed car worker:

> What most people see [in FSU] is somewhere to go for advice, somewhere to go and sort a problem out. I mean we've got somebody who will talk to you and they will help you, guide you in the right direction, how to get things done. The other part of FSU is [that] most . . . all the time they try to involve people in doing things for themselves . . . Like, they will probably say 'this is how you go about it' and give them the phone to phone up and talk to the person [concerned] or whatever, or that sort of thing. Also, they get people involved actually in the groups there. If there's a lone parent, . . . a single mother with a child . . . [who] comes into the area, and they come here, probably for advice, they will be given the advice they need but also steered towards one of the women's groups or the nursery campaign. (Interview, 1986)

A major recent development has been the establishment of the 'three estates residents board' to represent the interests of people living in the council estates of Pool Farm, Hawksley and Primrose Hill. An important effect has been to provide a forum whereby local people can discuss with the local authority the requirements of the community as a whole as opposed to the needs of specific individuals. The benefits of such an arrangement were described by the unemployed car worker just quoted, as follows:

> One of the first things . . . is . . . environmental services in the area, in the way that dustbins never get emptied properly . . . Also, we used to have skips . . . that [were] dropped about twice a week, took away, emptied and brought back. What's happened to them? . . . The other issue is street cleaning . . . We want street cleaners in this area . . . [and] . . . caretakers . . . Which they could easily do, with some of the money that they're

> wasting on trips abroad and all that. [They are] the sort of issues that we'll take up first and then carry on from there. (Interview, 1986)

It is fascinating, but not altogether surprising, to note that, although they are at different ends of the housing ladder in South-West Birmingham, the basic aspirations of the residents of Bournville and the outer fringe estates are similar: decent housing in a decent environment. Furthermore, although they might not care to have it expressed in this way, the spirit of the FSU activists in the 1980s is not too far removed from that of George Cadbury, the benevolent capitalist, in the 1890s. Cadbury's ideal was an active, involved, self-respecting individual within a caring community. Perhaps significantly, among the FSU's benefactors is the Edward Cadbury Trust. It is probably, however, sheer coincidence that the land upon which the three estates were laid out was donated to the city by Cadbury Bros Ltd decades previously in 1937.

Bournville Village Trust and the South Birmingham FSU encompass only a small minority of South West Birmingham's residents. A more typical point of contact for local tenants is the city council's neighbourhood office. Although, as Beuret and Stoker (1986) point out, this institution is intended to bring the council and its services closer to the people, two other functions of neighbourhood offices are mentioned by local activists. They are, first, that their existence allows local councillors to shrug off a significant part of their caseload and, secondly, that the offices take much of the direct pressure (for repairs, decorating, cleaning and so on) away from the town hall, containing it at the periphery. In the words of the manager of one neighbourhood office in South-West Birmingham:

> As soon as you start up the office you create a demand and build. 'If you have a local door someone will come and knock on it' was a cynical saying that went around. You get flattened by the demand. It's probably better than before [but] . . . There's no restructuring of the council power structure. (Interview, 1986)

Neighbourhood offices vary considerably in the range of services they offer and in the degree of commitment they give to their clients' interests. As one seventeen-year-old pregnant girl put it: 'Get yourself a good social worker. Some of 'ems bastards [but] get yourself a good social worker. That's how I got a place [to live] . . .'

The neighbourhood offices are serving a population under considerable stress. As the office manager quoted above said, 'if you're out working, you're more vulnerable to burglary. . . Basically everything gets nicked as soon as a flat's empty . . . People expect to be burgled four times a year . . . Usually it's the front door or windows smashed, and we spend a lot of time getting those fixed.' He was speaking of parts of Kings Norton, but even Bournville is not immune. A recent annual report of the Bournville Village Trust noted:

> 1985 was yet again a year in which the figures for crime were on the increase. Bournville is not alone in this trend which is repeated throughout the West Midlands generally, but this part of Birmingham does have the disadvantage, and the dubious distinction, of having not only the highest density of accommodation in the whole Region, but also, the highest burglary rate as well. (Bournville Village Trust, 1985).

7 Conclusion

In conclusion, the historical context of present troubles should be re-emphasised. For almost a century up until the 1960s, South-West Birmingham was an expanding frontier of opportunity, which encouraged an ethic of individualism in both labour and capital. This was tinged with the aggressive chauvinism given full voice by Joseph Chamberlain around the turn of the century. The other side of this coin, in apparent contraction, was the humane paternalism of Birmingham's 'civic gospel' and the Cadbury's Bournville Village. These diverse ideological strains lived easily side by side at times of relative affluence. Trade unionism flourished most successfully during the post-war boom, although its local ideological roots were shallow; membership was instrumental for the most part.

When the slump came, there was little local solidarity. Hard times undermined the unions and intensified the division between working families and welfare recipients. Apart from localized coping responses, the locality depended for support upon agencies organized at higher levels. At the regional level, the development of a spatial coalition has been tardy and conflict-ridden. Ironically, the major source of practical help to the locality – in the sense of providing jobs with medium-term security for a significant proportion of local residents – has been the very companies whose redundancy programmes attracted such adverse publicity in the late 1970s and early 1980s. Although both companies want to maximize good will locally, neither sees itself as a major local benefactor. However, closure of either the Bournville factory or the Longbridge plant would present an additional burden that the locality is very ill-equipped to meet.

Acknowledgement

The research upon which this paper was based was carried out as part of the South West Birmingham locality study within the ESRC-sponsored Changing Urban and Regional Systems Initiative (DO4250006). The interviews were carried out by Michael Rowlinson, Malcolm Maguire and Dennis Smith.

References: Chapter 5

Beuret, K. and Stoker, G. 1986. 'The Labour Party and neighbourhood decentralisation: flirtation or commitment?', *Critical Social Policy*, **6** (2), 4–22.

Birmingham Chamber of Industry and Commerce. 1983. *Reversing structural decline in the West Midlands*. Birmingham: BCIC.

Birmingham City Council. 1981. *Report of Housing Committee*. Birmingham: BCC.

Bournville Village Trust. 1985. *Annual Report for 1985*. Birmingham: Bournville Village Trust.

Cartland, B. 1942. *Ronald Cartland*. London: Collins.

Church, R. 1979. *Herbert Austin. The British motor car industry to 1941*. London: Europa Publications.

Dellheim, C. 1987. 'The creation of a company culture: Cadburys 1861–1931.' *American Historical Review*, **92** (1), 13–44.

Hensloe, P. 1984. *Ninety years on. An account of the Bournville Village Trust*. Birmingham: Bournville Village Trust.

Liggins, D. 1977. 'The changing role of the West Midlands region in the national economy.' In F. Joyce (ed.) *Metropolitan development and change*. London: Teakfield.

Morgan, K. 1985. 'Regional regeneration in Britain: the territorial imperative and the Conservative state,' *Political Studies*, **33**, 560–77.

Rex, J. and Moore, R. 1967. *Race, community and conflict. A study of Sparkbrook*. Oxford: Oxford University Press.

South Birmingham Health Authority. 1986. *Profile: health, finance, activity, population*. Birmingham: South Birmingham Health Authority.

Spencer, K. *et al*. 1986. *Crisis in the industrial heartland. A study of the West Midlands*. Oxford: Clarendon Press.

Sutcliffe, A. and Smith, R. 1974. *Birmingham 1939–70*. Oxford: Oxford University Press.

West Midlands County Council. 1981. *Regional industrial development*. Birmingham: WMCC.

West Midlands TUC. 1982. *Our future*. Birmingham: TUC.

6

Regency icons: marketing Cheltenham's built environment

HARRY COWEN

1 Introduction

Cheltenham, situated on the boundaries of the West Midlands and South-West England, yet only ninety minutes by road from London, has been subject to considerable economic transformations in recent decades. But the most powerful policy impacts relate to its built environment and the continuing strengths of its more conservative traditions. The local state's manipulation of the built environment has been particularly significant. Its reshaping of the prestige social imagery of the leisured Victorian town centre has been central to the area's economic restructuring. This has added to the historical accumulated layers of investment and an educated labour force within a 'Thatcherite' restructuring of the UK economy, characterized by disinvestment in the traditional industrial conurbations. Cheltenham provides a classic example of place marketing in the 1980s, where the capitalization of heritage locally has coincided with an era of national heritage building policies and a museum culture that has exalted the value of age. This coincidence has been fortuitously bolstered by an added symmetry between Cheltenham's traditional military defence strategic and manufacturing base and the Conservative Government's 'defence of the realm' policies.

This chapter commences with a brief historical outline of Cheltenham's traditional economic development policies from 1918 until the end of the 1950s. The second section focuses upon Cheltenham's office policies and consequent reproduction of the built environment and traditional locality image, considering the respective roles of finance capital, the conservation lobby, and town planners, in response to the wider forces of economic decentralization. The third section considers the other spearhead of Cheltenham's contemporary policies – the transformation of much of its centre into the county's prime retailing area. The fourth section looks at residential development policies, and the fifth at the planners' tourist strategy. Then the

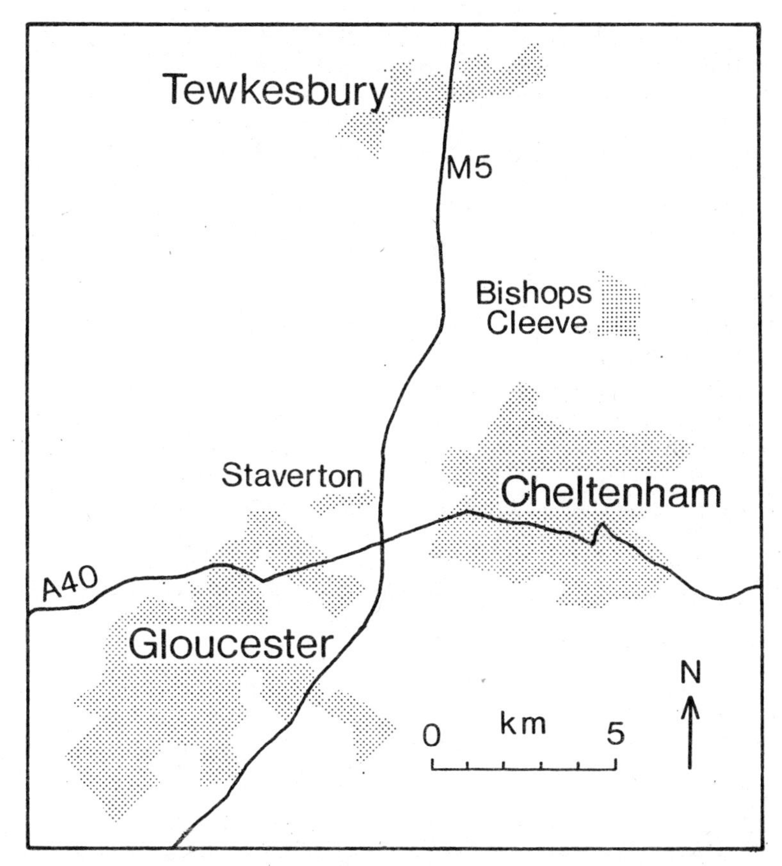

Figure 6.1 Location of Cheltenham.

politics of these policies are considered. Finally, certain conclusions are drawn regarding the relationship between the locality's policies, its politics and national economic restructuring.

2 History of economic development

An assessment of the achievements of the local state in the development of the Cheltenham economy serves to highlight the contrast between its positive economic and council housing activity throughout the earlier years of industrialization from 1919, and its activity during the 1980s when the local authority did not seek to attract manufacturing industry, and demonstrated little interest in maintaining levels of council house provision. Council house building had been conspicuous during the inter-war years, partly in response to national housing legislation and the raising of housing standards but also in reaction to industrial pressures, particularly to the rapid expansion of local aircraft companies during the First World War. Thus Cheltenham Borough Council purchased land in the St Mark's district for 650 council houses, and largely earmarked them for the key workers required by Smiths Industries. Virtually 50 per cent of all housing built in Cheltenham during the inter-war years was built by the local authority. Between 1935 and 1939, with more engineering factories locating in Cheltenham, the local council doubled its housing stock to a total of 2000 (Hurley, 1979, p. 127).

In addition to housebuilding, the council also ran a vigorous campaign between the two world wars, using nationally circulated posters to attract light engineering firms, offering well-located factory sites (Blake and Beacham, 1984). Other attractions included a pool of trained engineers plus the relatively low wage levels conditioned by the area's long-standing agricultural hinterland and its domestic service industries (Hurley, 1979, p. 117). During the 1920s, for example, more than seventy in every 1000 females in Cheltenham worked in domestic service, compared with fewer than thirty in Gloucester (Population Census, 1921).

After the Second World War, the local authority again adopted an explicit policy of attracting the most desirable industries, as illustrated by a Chamber of Commerce guide: 'Should you control a business that entails chimney stacks and ugly dumps then put this aside.' (Cited in Hurley, 1979, p. 149). Less appealing establishments, such as Birds Eye Walls, requiring mainly unskilled labour, were dissuaded from locating in Cheltenham and directed to Gloucester. Cheltenham also accumulated specialist prototype and precision engineering firms. To this extent, the locality was undeniably pro-active in shaping the post-war economy according to a pre-conceived social pattern. By contrast, the move of the Government Communication Headquarters (GCHQ) to Cheltenham in the early 1950s provides an example of national state direction and funding determining local authority housing activity, when the government financed Cheltenham's construction of 500 houses and 500 flats especially for GCHQ employees. Later, in the 1970s, private housebuilding increased greatly in the Benhall area adjacent to the second GCHQ site.

Cheltenham's economy experienced a fundamental restructuring in the 1960s, with the building up of a producer services sector to compensate for the declining employment opportunities in direct aircraft manufacturing, which had occurred in the late 1950s (Livingstone, 1987a). Diversification was initially stimulated by the national policies aimed at decentralizing offices from London, through which a substantial insurance industry developed in Cheltenham (McNab, 1987). The wave of office decentralization continued throughout the 1970s, affecting private companies and state agencies in the educational and environmental sectors.

The restructuring of Cheltenham's industry towards producer services, well advanced by the 1970s, continued into the 1980s, not least because of advantages conferred by locational features and the quality of the built environment. Insurance, banking and related activities employed 9.8 per cent of the Cheltenham labour force by 1981, substantially above the 6.2 per cent national average. Its importance had grown by 1984, with the proportion standing at 10.5 per cent (figures from NOMIS (National On-Line Manpower Information Service)).

Cheltenham's growth, like Swindon's has partly resulted from government decisions to decentralize its own departments. Both areas also benefited from the unsuccessful efforts by government to steer private offices to the industrial regions of the country such as Merseyside and Teesside. In addition to such national factors, local authority initiatives were apparent, through the attraction of light engineering for example. Such policies, albeit for differing reasons, gained the support locally of employers, organized labour and the three mainstream political parties – Conservative, Labour and Liberal.

By 1980 Cheltenham had profited from the overall capitalist restructuring that had occurred (Massey, 1984; Hall, 1987). Whereas professional and managerial employees accounted for 29 per cent of Cheltenham's total employment at the beginning of the 1970s, they represented almost 35 per cent in the early 1980s (compared with 27 per cent in Great Britain as a whole). Cheltenham's proportion of employers, managers and intermediate non-manual workers is above and growing faster than the Great Britain figure. Correspondingly, the Cheltenham population is better qualified than the national population. Likewise, compared with a national employment growth of 1.1 per cent between 1971 and 1981, Cheltenham's working population grew by 3.4 per cent. By contrast, Cheltenham registered the lowest rate of growth of unemployment in any area between 1972 and 1982 (Cooke and Morgan, 1985).

The role of the national state as a stimulus in the local economy has throughout the century constituted an important feature of Cheltenham's growth. This includes its strategic decisions regarding the defence and aircraft related industry (Cowen, 1987) and intelligence-gathering activity (Livingstone, 1987b), and also its office decentralization policies. Nevertheless, as this chapter demonstrates, the local state and local political forces have had an active role in guiding the locality's economic trajectory, not least through policies affecting the built environment.

3 Office development policies: reproduction of the built environment

In this section I shall explore how office development was affected by market forces, and shall also focus on the considerable role played by Cheltenham's local policies and strategies in capitalizing upon both natural and built environmental resources. The Cotswolds backcloth to Cheltenham has greatly influenced its long-term development. It presents an ambience or image that evokes retirement and leisure, rather than industry. The locality's broader environment is important in any attempt to understand the area's contemporary political economy.

Built environment and planned landscaping have been central to the communication of the locality's image. Much of Cheltenham's building stock was produced for the dominant sections of the local economic and social structure. In the eighteenth century this was fundamentally the Georgian aristocracy, and by the middle of the nineteenth century it comprised the gentry, leisured ex-colonial officials and army officers, and the rising bourgeoisie (Hurley, 1979, p. 112). The Montpellier area's antique shops and statues of mock Greek antiquity, the Rotunda, the Pittville parkland and Pump Room have been retained as familiar symbolic as well as spatial landmarks, and the vast expanses of public school sports fields and architecture are stressed by the Cheltenham Borough Local Plan (1985a). More than one-third of Cheltenham's open space is now owned by private schools or clubs (Cheltenham Borough Council, 1985b, p. 27). So Cheltenham's inner area comprises a built resource where historical and ideological symbols have been refurbished and marketed by the town's economic and physical planning policies; the latter have represented the interests of both finance capital and local Conservative political interests.

A series of conflicts about the exact form of land and property planning developments reflected both the post-war Conservative ascendancy of the locality's politics and the fact that power oscillated between contending factions of the Conservatives, finance-bourgeoisie and property interests, on the one hand, and the preservationist/conservationists on the other, who were opposed to industrial development and new building in the urban centre.

As we shall see, this situation substantially explains why the preservation of the old Spa ambience has coincided with Cheltenham's renewed economic growth and a set of central government policies simultaneously reinforcing market individualism and social elitism. The Cheltenham environment has served as a pole of attraction for both capital and labour, and has communicated a series of conservative ideological and political symbols orchestrated by the local state. It has been manipulated in the course of the shaping and re-shaping of the area's economic and social structure, reinforcing relationships between environmental policy and place (this is echoed in Chapter 7 on Lancaster, 'where place is understood in terms of its very buildings'. (p. 201)). I shall now consider the main developments in the 1960s and 1970s.

The office decentralization policy of the mid-1960s benefited Cheltenham, because offices could be freely located beyond the restricted development zone around London. A Tory council dominated by property interests acted expeditiously. Between 1967 and 1976 Cheltenham, like Swindon, became a centre for speculative office development. During this time, office floorspace expanded locally by 0.75 million sq. ft. (Shapira, 1977, p. 49). The local authority concentrated on attracting prestigious institutions in the private finance capital sector, offering information on desirable locations and rate inducements. Some of the resulting blocks were sufficiently ugly architecturally for the town's redevelopment to be described in *The Rape of Britain* as follows:

> The spacious elegance of Cheltenham is under threat of gradual erosion. Large villas are empty and derelict and the stucco fronts of the houses and terraces are crumbling and in need of constant maintenance. New developments are too often unresponsive to the nature of the town . . . New developments are particularly damaging when they are out of scale, and the headquarters building for the Eagle Star Insurance Company towers over the town and ruins the view over a wide area, because of the sheer bulk and size. (Amery and Cruickshank, 1975, p. 55).

Although office development was directed to selected sites after 1968, the opposition to the predominance of office and speculative development in the town was considerable, and ran across party lines. A 1966 plan to redevelop totally Cheltenham's town centre (and demolish much of the Regency building stock) split the Conservatives. The Liberals favoured conservation (Hurley, 1979, p. 164), and the Labour Party generally supported the plan. A conservation lobby formed around the Cheltenham Society, comprising professional and managerial middle-class residents. Opposition started immediately the plan was published, and the lobby's strength during the 1967 public enquiry led to its eventual rejection.

By the 1970s the council turned instead to using the town's Regency building stock for conversion into prestige offices. Through a panoply of council policies, including rates waivers and permissive planning, and because of its attractive environment, Cheltenham had appealed to insurance multinationals, such as Eagle Star and Mercantile & General, government departments and quangos, and the head offices of industrial companies such as Gulf Oil and Kraft Foods. Local council officials enthusiastically launched promotions and receptions for potential prestige clients. The offer of a celebrated Regency heritage building to be renovated (now the John Dower House) was inducement enough for the Countryside Commission to leave London in 1970 (Interview – Chairman, Countryside Commission). Throughout this period the local state pursued a policy of Regency building conservation which matched and indeed assisted its office development policy:

> by preserving the town's elegant facade, which helped to secure an elite position for Cheltenham as a headquarters town for the institutions themselves. The attractiveness and image of the environment were important factors in giving the town status with the institutions, and the inflow of investment capital would surely generate monies for conservation (Hurley, 1979, p. 168).

Given the concern for the marketing of image, grants were offered for the conservation of facades rather than for interiors. This initially applied to Regency buildings that were being converted into offices, but was later extended to the Regency housing stock as a whole. At the same time, conservation pressures from both right and left pushed for a cessation of new office building. In 1973–4 the conservationist group and the Cheltenham Housing Campaign (basically comprising student activists and left-wing organizations – Labour Party, Communist Party, International Socialists) opposed office development. However, there was no formal coalition between the traditional conservationists and the left-oriented housing campaign. Whereas the latter campaign related office development to housing shortages, the conservation groups linked it to environmental damage. Thus, the council's office development policy was dropped in 1974 and a conservation policy adopted unequivocally. This involved the wholesale rehabilitation of the Regency housing stock.

However, finance and property interests were still involved in major developments, but these entailed building in harmony with the Regency environment – for instance, the Royal Scotland Trust's Royscot House which, according to Britannia Developments' glossy brochure, evinces:

> THE TRADITIONAL ASPECT: WHERE ESTABLISHED PRINCIPLES REFLECT CORPORATE VALUES:
>
> . . . carefully and imaginatively designed – in conjunction with the Fine Arts Commission – to blend and harmonise with the imposing Municipal Terrace; . . .
>
> THE ASPECTS OF QUALITY:
>
> . . . there are few more attractive places in the U.K. for international clients to visit . . .
>
> THE ASPECT OF PROGRESS:
>
> Where future success reflects a progressive outlook . . . Moreover, the building also gells with:
>
> the historic buildings of the world-renowned Cheltenham Ladies College diagonally opposite.

In another variant of modernized conservation, the Cheltenham and Gloucester Building Society's new main offices have occupied a listed building that has been refurbished in mock Georgian style. This architect-planning compromise between market forces and controlled conservation (or marriage of convenience between property capital and planning constraint) has not passed unremarked by the Cheltenham population. 'A Cheltenham native' wrote to the local paper:

> I have become increasingly distressed that, all over this beautiful town, buildings of character and mellow stone are being torn down and replaced with Toytown mock-ups, complete, in many instances, with 'eyeless' windows.
>
> This Borough Council have presided over this desecration as they have grovelled to encourage every greedy commercial enterprise to this town, more interested in money than preservation . . . For how much longer must we suffer the vulgarisation of our lovely town? (*Gloucestershire Echo*, 26 January 1986).

A recent large-scale development has involved the centrally located St James Station site, which has remained a recurrent planning issue. The lack of any formal plan for this area meant that the issue remained unresolved for a quarter of a century. A planned deal with the Cheltenham and Gloucester Building Society in 1987 collapsed, with the Cheltenham Borough Council holding firm over the question of the price for the site, a self-declared testimony to the relative power and autonomy of the local authority. In the event, the rebuffed Building Society relocated its headquarters in Gloucester. The St James site now houses both the Polytechnic Central Admissions System and the Universities Central Council on Admissions together in a five-storey, 58 000 sq. ft. office building capable of housing UCCA's major new computer system.

The following discussion on retailing investigates a further example of the inter-meshing of market forces, a growing middle-class 'designer' consumerism and the mediation of the planning machinery.

4 Retailing policy: consumerism and property capital

This section traces the ways in which the built environment has been used to promote major retailing developments and property capital interests. Although in the 1960s and 1970s, with the arrival of many middle-class professional workers, Cheltenham had experienced much speculative housing and office development, the range of retailing facilities still failed to match that of neighbouring Gloucester. Cheltenham's retailing divided into two discrete shopping quarters: the High Street, with some of the national multiples, and the fashionable Promenade, which has continued to project the image of a shopping mall for the rich. But retailing developments in the 1980s have altered this situation. The Council has sought to promote the town as a high quality regional shopping centre.

Three distinct types of retailing can be distinguished: a renewed shopping centre in the city centre; a high-quality speciality shopping district; and edge-of-town retailing (O'Brien and Harris, 1986; 1988). The shopping centre has been crucial for Cheltenham. Regent Arcade, an architecturally celebrated development in the central Conservation Area (Herzberg, 1985) is a prime example, bridging the Promenade and the High Street inside the rehabilitated classical Plough Hotel. This is an imaginative if controversial response from a commerce–developer–planner consortium to the town planners' attempts, in the face of prevailing pressures from commercial capital, to save Cheltenham's historic centre, through sponsorship of an architectural competition. This development was in line with the Interim District Plan's fundamental goals which were to protect the town's central shopping facilities from excessive competition by edge-of-town developments; to improve central shopping facilities by attracting new multiples; to increase Cheltenham's popularity as a tourist centre for visiting the Cotswolds, the South West of England and other adjacent areas; and to uplift the town's architectural character by the provision of a compact, managed shopping centre whose design was not antithetical to the predominant local Regency architecture (Cheltenham Borough Council, 1978).

Cheltenham's second type of retailing development has been expressly dovetailed with local rehabilitation and conservation policies designed to promote expensive speciality shopping in the Montpellier area. The neighbourhood had deteriorated through neglect and the

Spa's general loss of competitive position vis-à-vis other spa towns at the turn of the century (Pakenham, 1971, pp. 138–51). Its architectural revival has been greatly facilitated by a central government special environmental funding scheme in operation since the early 1970s, in concert with the proactive conservationist local authority. Designated as a conservation priority area, Montpellier became a direct beneficiary of Britain's urban architectural and environmental conservation movements. Grants of up to half of the rehabilitation costs have been distributed for enhancing the Regency environment and fabric, financed equally by Cheltenham Council and the Historic Buildings and Monuments Commission. New speciality firms have moved in, taking advantage of renovation grants and the supply of vacant, cheap buildings to service the growing local and sub-regional professional consumer market. At a later stage, a private development, Montpellier Courtyard, a small exclusive shopping mall of post-modernist design was stylistically blended with the existing Regency vernacular. A plethora of wine-bars and intimate restaurants have completed a retailing district which bears comparison with those in other historic towns, for example Bath and York, which have aimed to profit from their own heritage building stock.

Finally, a third type of development illustrates the strains placed upon a heritage locality enveloped by dominant market forces and the decentralizing tendencies of basic product shopping. With the extension of the suburbs in the wider area around Cheltenham, there has been a pronounced trend (as elsewhere) for retail suppliers to re-site their commercial operations on the town's fringes. Several local authorities have allowed such commercial strategies. But Cheltenham planners' hostility reflects the conscious avoidance of any threat to Cheltenham's own comprehensive central shopping strategy, by maintaining its exclusive characteristics and thereby its current position as the regional shopping centre.

Without doubt the retailing developments owe their momentum to the ruling politics of the market. But also, the strength of the local conservation lobby has meant that the planners have avoided earlier policies of comprehensive demolition.

5 Residential development policy

The Local Plan eventually adopted in the 1980s has emphasized the conservationist theme of preserving the character of Cheltenham through its open spaces and architecture:

> (The) character of the town is recognised as having national as well as local significance. The Plan seeks to maintain this character in a number of ways: by restricting the future growth of the town, by protecting and enhancing important open spaces and historic buildings, by improving areas or sites which are run down or unattractive, and by endeavouring to ensure that any new development is of a high standard of design and harmonises with existing buildings (Cheltenham Borough Council, 1985a, p. 12).

Most of the town's inner core is designated as a Conservation Area. By 1985 two thousand buildings had been listed as having special architectural or historical merit. A total of £9.5m in grants had been channelled into the town's private Regency housing stock for house improvement and renovation. In 1988 Cheltenham was among the ten highest spenders on historic buildings in an English Heritage Monitor survey of borough and district councils. The massive Regency housing rehabilitation programme undertaken by the local council in the 1980s produced effects that have been of especial benefit to the higher income earners, largely as a result of the up-grading of the inner area housing stock and the consequent conversions into 'single professional, up-market' apartments. Large grants to owner-occupiers living in Regency properties in designated neighbourhoods such as Lansdown, adjacent to fashionable Montpellier, have constituted a considerable subsidy to the private housing market.

The money spent by the Council on rehabilitation has been exceptional. Receipts from central government grants ensured that Cheltenham's gross expenditure on improvements averaged some £89 000 per head, compared with £55 000 in Gloucester and £62 000 in England as a whole (Interview with CBC Deputy Treasurer, 1986). A secondary effect has been to raise house-price levels, making the Regency housing and apartment market more than ever a 'high income' middle-class market. In the words of the Secretary to the Cheltenham Association of Real Estate Agents: 'What the private housing market is doing now is undoubtedly being influenced by the local authority.'

Accompanying the Thatcherite 'privatization' of public sector services has been the rapid development of sheltered housing schemes and private nursing homes for the elderly in the Cheltenham area. Between 1980 and 1986, companies built 290 units, mainly for in-migrant elderly persons, and 48 private and voluntary homes, converted from the vast Regency and Edwardian properties ringing the town, housed almost 950 old people (Cheltenham Borough Council, 1986a).

But the rehabilitation policy has had a wider import. The local planners have remained fully conscious of the economic 'spin-off' obtainable from the nurturing of a rather rarified view of local history through the built environment.

> In keeping with the town's history, the architecture, along with other attractions, has resulted in the development of a thriving tourist industry, and also attracted the large scale office development in the 1960's and 1970's which brought many firms to the town from other parts of the country. (Cheltenham Borough Council, 1985a, p. 12).

A focus upon 'prestige' symbolism and the re-creation of a 'ruling class' heritage, most transparently evoked by the public schools and Regency mansions, but one that carefully masks the town's artisan and working-class history (Ashton, 1983), has attracted finance houses, investment companies and building societies during an epoch in which 'history' and 'heritage' are indispensable municipal marketing icons, irrespective of their location (Hewison, 1987). Local examples include Hill Samuel Investments Ltd, accommodated at 'Royal House', in a quarter of the town overshadowed by the Cheltenham Ladies College; and the Chelsea Building Society, which opened its national administrative headquarters in a rehabilitated Regency mansion close to the Cheltenham Boys College. In 1986, London-based developers took over Montpellier House, a former Ladies College boarding house dating from the 1880s, for conversion into flats costing more than £1m each. This rehabilitation of one of Cheltenham's largest residential houses included interiors fitted with antique furniture to blend with the period of the original house. The Sales Manager of the show flat stated:

> The Montpellier House project has been very exciting from day one, but this is the icing on the cake. It's so exquisite, it's rather like showing people round a stately home. (*Cheltenham Source*, 24 July 1986, p. 7).

In general, capitalizing upon middle-class values in the market for ancient properties has been much in evidence. Appropriately, the housebuilding developers Bovis Homes and Westbury Homes occupy vast, elegantly refurbished mansions along the town's thoroughfare towards Gloucester. But, of course, the value of their location is not just symbolic. Residential 'period house' properties are becoming the most desirable on the housing market, attracting four times as many buyers as modern properties. (*Independent*, 12 September 1987). Until fairly recently many had fallen into a state of disrepair and could be purchased cheaply. But, with a heightening awareness of the value of historic property of architectural merit, prices are rapidly increasing. As noted in a recent issue of *Landscape*:

> [It] is not for nothing that many estate agents in towns with Georgian centres are to be found elegantly installed in one of the finer listed town-houses (O'Reilly, 1987).

Politically, the forces of private property capital, the middle-class conservation lobby and the council's planning policies would seem to have combined to a degree that has left working-class council tenant interests virtually powerless.

6 The tourism strategy

Surprisingly, in the light of the previous section, Cheltenham's tourism strategy has been somewhat hesitant, although the conflict between the pure conservationists and the forces of capital using conservation offers an indication of why this should be so. In fact, there has been the projection of sense of place rather than the quest for a tourist industry as such. Nevertheless, the pressure on local authorities to accumulate revenue, against a background of extensive economic restructuring and fading central government financial support, has prompted councils throughout Gloucestershire to plan not so much for an increase in the number of tourists but to maintain existing levels. In 1987, the combined local authorities of the county launched a tourist promotion strategy, a £30 000 campaign aimed at overseas visitors. A county tourist officer was appointed for the first time and a glossy brochure marketed Gloucestershire as 'England at its Best'. The county council was one of the first in England to undertake a joint promotion of this kind. Clearly, the selling of the Cotswolds environment and, indeed, a 'way of life' as a total package simultaneously draws the tourists to both town and country, and at the same time shares any negative environmental impacts and reduces the investment costs for any single authority.

Politically, the more conservative forces in Cheltenham have tended to oppose the spending required for a viable tourist policy, and have maintained an emphasis upon the preservation of the built environment. Concern was expressed by the Liberals, the new ruling group on the Council in 1986, at the meagreness of the Borough's tourist publicity budget. But the Conservatives were against increasing spending without an 'appropriate' cutting of other services. Cheltenham planners had produced a report that proposed increased expenditure because of changing national tourist patterns: for example, long-distance day-tripping to places such as Cheltenham or the concomitant decline of seaside holiday resorts like Morecambe. Escalating competition from many other local authorities had now changed the market for the style of product Cheltenham felt able to promote. One such authority is Lancaster, recently marked out for a Tourist Development Action Programme (see Chapter 7). Cheltenham's strategy document noted, almost in surprise:

> Even towns such as Bradford and Wigan, which have traditionally been regarded as non-tourist towns, are marketing their particular attractions positively and are investing in facilities for visitors with some considerable success. (Cheltenham Borough Council, 1986b, p. 6).

A nearby competitor seen as offering tourist resources was Gloucester with its cathedral, docks and the National Waterways Museum. In contrast, Cheltenham was deemed to have inadequately exploited its singular attributes and image marketing in relation to tourism: namely, the cultivation of its environmental and cultural ambience. The council planners advanced an overall strategy for improved marketing. To increase the number of tourists, the strategy, largely founded upon the selling of Cheltenham as 'the finest complete Regency town in the country', proffered a set of opportunities, embracing its gardens and parks, and 'holiday education'. Nevertheless, opposition emanated from both the conservationists and the Conservatives. Responding to the Tourism Strategy, the Cheltenham Civic Society stated that a superfluity of visitors would ruin the town's character. Opposition from the Conservatives included the MP, who argued that the lack of large hotels prohibited any ambitious tourist policy.

So a direct jobs-related tourist economy has been less prominent than strategies to develop the shopping fabric and the continued use of heritage buildings for attracting financial and public sector employers. Image marketing has constituted a major strand of these policies. The conservation of the built environment has been pivotal to the project. To this extent, planning policies have played a key role in Cheltenham's economic development. A consistently high level of heritage improvement grant money has been channelled into retailing and property schemes, as well as the rehabilitation of the Regency housing stock to accommodate sections of in-migrant, more affluent members of the service class. This offers ample illustration of the locally inspired policy of marketing place, history and architecture in order to derive full benefit from contemporary economic restructuring and central government strategies.

An analysis follows of the politics of such policies and the relationships between local policies, politics and national economic restructuring.

7 Policy and politics

Four key spheres of policy in Cheltenham have been investigated: office development policies; retailing; residential development; tourism. In each case, a series of influences was at work: private market forces, local council and central government policies, and the contending interests of particular local social groups. More recent projects have been stimulated by the overall political climate, national policies rewarding the private sector and the restructured spatial division of labour, but they have also been moulded by the locality's political history and contemporary political re-alignments. The latter have been graphically expressed in the arena of land use and the built environment. The conservation versus development theme has monopolized the locality's political life in recent years. Disagreements about this issue cut across party lines, but at a time when Conservatives at both county and district levels were totally dominant. In this respect, the differences reflected the variable elements of local conservatism.

As outlined earlier in the chapter, the Conservatives have represented both finance capital and landed property interests (especially effective in the town's 1960s policies geared to new development) and the 'preservationists', who resisted modern schemes. The preservationist group has gathered support from diverse political formations because of a shared approach to the environment. These interests have included the political environmentalists, anti-office-block pressure groups and even socialist groups actively campaigning for decent working-class housing. Notwithstanding the correspondence of interests, there is no evidence of formal coalition politics (as, for example, in Thanet). But there have been periods where forms of alliance were feasible, usually based around the conservation-oriented Tories. In effect, environmentalist pressure groups have tended to comprise professional workers turning away from what they perceive as less commendable forms of market capitalism. Thus, the Cheltenham Society was pro-conservation, ostensibly non-political, and comprised ageing intellectuals and professionals employed by local colleges, intent on preserving the built fabric and images of the past and countering the encroachments of industrial capital and discordant building design.

The local Green movement has also exerted an influence on the locality's political and economic policies. The presence in the 1970s of a predominantly middle-class Friends of the Earth group was an outgrowth of the broader environmentalist movement of the 1960s. Friends of the Earth was replaced in the 1980s in Cheltenham by the Green Party, which sought to place the environment on the electoral agenda. Ironically, the party's platform was 'hi-jacked' by the Liberals, who realized the political advantage of pursuing policies in line with the thinking of many of the area's recent professional in-migrants. Because the Liberals had also been less involved with big property capital, they had been less divided than the Conservatives on the question of new development. Furthermore, Cheltenham has had a strong independent liberal tradition.

Liberal local electoral victories in the 1980s meant that the voice of the property owners among the Conservatives was slightly muted, with the Liberals' relatively greater emphasis upon public projects, direct job schemes and preservation of the environment. However, this emphasis can scarcely be described as a new policy direction. The power of national finance capital, and the fact that local business interests have still not been displaced has led to a fusion of the diverse conservative interests, assisted by the activities of the town planners. The economic and conservationist interests have come together to support town centre redevelopment within a conservationist framework of rehabilitation and the conversion of historic and civic buildings. This policy was helped by the 1974 local government reorganization, which had ensured that town planning powers, previously vested in the county council, augmented the borough council's ability to advance an integrated strategy for the locality's economy, its environment and housing stock, outlined in a forceful, conservationist-cum-retailing town plan in 1981.

Finally, the effectiveness of the Labour Party in this climate has been impaired. Indeed the heavy defeats they have suffered in recent elections (in parallel with national trends) have silenced the local Party's calls for fresh industrial employment opportunities. Its share of the parliamentary constituency vote dropped from 27.7 to 7.6 per cent between 1974 and 1987. Its vote had peaked in 1966, a time when the rise of the finance sector in the area was only in its early stages.

The locality's subsequent industrial restructuring, accompanied by a growing predominance of the professional service class, has marginalized Labour (in contrast to the situation in Lancaster).

8 Conclusion

The fate of local economies such as Cheltenham has depended particularly upon national policies and developments. Between the wars, the local authority was able to play a role in creating a manufacturing base, at a time when industrial production was the most viable form of economic activity. More recently, in the light of broader restructuring processes, local diversification strategies have been deemed necessary, in a climate favouring service industries and the knowledge skills of an educated service class. To this extent, locally pro-active policies are again in evidence. Features such as relative proximity to London and environmental quality take on a new dimension with the systematic marketing of place. Local policies and strategies have, in Cheltenham's case, consisted of two projects: the manipulation and marketing of the built environment, and the development of high-class shopping schemes.

Since the late 1960s and the successful diversification towards producer services and finance capital, the strongest political forces have aligned themselves with policies communicating an anti-industrial, professional middle-class image for prestige urban developments. This has been even more so with recent electoral changes (the first real local political party shifts in council power since the beginnings of this century) and a greater representation from those with less of a personal stake in property.

National restructuring processes in Britain in the 1980s have destroyed economic opportunities in traditional industrial localities, but they have generated new rounds of investment in specific types of locality that possess distinct characteristics, such as a heritage building stock. These characteristics have been capitalized upon by local planning, housing and environmental policies, which have facilitated diversification of the economic structure, but to the undoubted advantage of the middle class. To this extent, the locality's built resource has proven to be a key determinant in the new climate of local economic initiatives.

Acknowledgements

I should like to thank the Cheltenham Locality Research team: Ian Livingstone, Andy McNab, Steve Harrison, Bryan Jerrard and Lawrie Howes for earlier assistance; and also planners Paul Fry and David Hunt.

References: Chapter 6

Amery, C. and Cruickshank, D. 1975. *The Rape of Britain*. London: Elek.

Ashton, O. 1983. 'Clerical control and radical responses in Cheltenham Spa 1838–1848.' *Midland History*, **18**, 121–47.

Blake, S. and Beacham, R. 1984. *The book of Cheltenham*. Northampton: Barracuda Books.

Cheltenham Borough Council. 1978. *Interim District Plan*. Cheltenham: CBC.

Cheltenham Borough Council. 1985a. *Cheltenham Borough Local Plan*. Cheltenham: CBC.

Cheltenham Borough Council. 1985b. *Cheltenham Borough Local Plan: Report of a public local inquiry into objections to the plan*. Cheltenham: CBC.

Cheltenham Borough Council. 1986a. *Housing: monitoring of progress and policy 1985–86*. Cheltenham: CBC.

Cheltenham Borough Council. 1986b. *Regency Cheltenham Spa: Garden Town of England: Centre For the Cotswolds: a strategy for tourism*. Cheltenham: CBC.

Cooke, P. and Morgan, K. 1985. *Flexibility and the new restructuring: locality and industry in the 1980s*. Papers in Planning Research 94. Cardiff: Department of Town Planning, UWIST.
Cowen, H. 1987. *Defence engineering in the Cheltenham locality*. Cheltenham Locality Study. Working Paper 3. Gloucester: School of Environmental Studies, GlosCAT.
Hall, P. 1987. 'Flight to the Green,' *New Society*, 9 January, 9–11.
Herzberg, H. 1985. 'Regent Arcade by Dyer Associates: a new Promenade for Cheltenham.' *The Architects' Journal*. **181** (20), 15 May, 51–69.
Hewison, R. 1987. *The heritage industry: Britain in a climate of decline*. London: Methuen.
Hurley, J. 1979. 'Capital, state and housing conflict: a study of housing in Cheltenham.' Unpublished PhD thesis. Coventry: University of Warwick.
Livingstone, I. 1987a. *Restructuring and the Cheltenham economy: 1971–1981*, Cheltenham Locality Study. Working Paper 1. Gloucester: School of Environmental Studies, GlosCAT.
Livingstone, I. 1987b. *Working for the state*. Cheltenham Locality Study. Working Paper 5. Gloucester: School of Environmental Studies, GlosCAT.
McNab, A. 1987. *Finance and Insurance Industries in Cheltenham*. Cheltenham Locality Study. Working Paper 4. Gloucester: School of Environmental Studies, GlosCAT.
Massey, D. 1984. *Spatial divisions of labour*. London: Macmillan.
O'Brien, L. and Harris, F. 1986. *The character of shopping in the 1980s*. Occasional Paper, Geography Department. Cheltenham: College of St Paul and St Mary.
O'Brien, L. and Harris, F. 1988. 'The changing face of the town.' *Geographical Magazine*, December.
O'Reilly, E.-L. 1987. 'The cheapest Georgian townhouse.' *Landscape*, November, 19.
Pakenham, S. 1971. *Cheltenham: a biography*. London: Macmillan.
Shapira, P. 1977. 'The housing problem. Housing policy and the state.' Unpublished Diploma dissertation. Gloucester: Department of Town and Country Planning, Gloucestershire College of Art and Design.

7

Lancaster: small firms, tourism and the 'locality'

JOHN URRY

1 Introduction

The Lancaster area, with a population of about 125 000, is on the edge of the Lake District and is tucked into the north-west corner of Lancashire. Its development was peripheral to the core centres of nineteenth-century industrial growth in Lancashire, which were based on cotton spinning and weaving. It had been a prosperous port for the Atlantic trade in the eighteenth century but that had declined by the end of the century. Only later in the nineteenth century did Lancaster's fortunes revive with the development of production of the textile-related products, linoleum and oil-cloth. Morecambe, five miles away on the coast (and part of the same local authority since 1974) developed into a working-class holiday resort much later than many of the major resorts such as Blackpool and never attracted very large numbers of visitors from Lancashire itself.

By about 1920 Lancaster had become the major British centre for the production of linoleum and Morecambe had developed into the quintessential working-class resort, serving particularly the West Riding of Yorkshire (it became known as Bradford-by-the-Sea). New firms, especially in artificial fibres, were attracted in the 1920s, and there was fairly low unemployment, substantial in-migration and population growth right through to the early 1960s. In the post-war period, there was considerable investment in chemicals, oil-refining, artificial fibres, plastic-coated goods, and the tourist infrastructure. Even as late as 1964, local employers were complaining that there was more than full employment and that they were unable to recruit the labour required (Fulcher, Rhodes and Taylor, 1966, pp. 40–1). By the later 1950s Lancaster had developed an industrial structure highly favourable for future economic growth (Fothergill and Gudgin, 1979, p. 169), and the tourist industry in Morecambe was flourishing, with record numbers of visitors (Urry, 1987a).

Within a few years the situation had changed dramatically. The

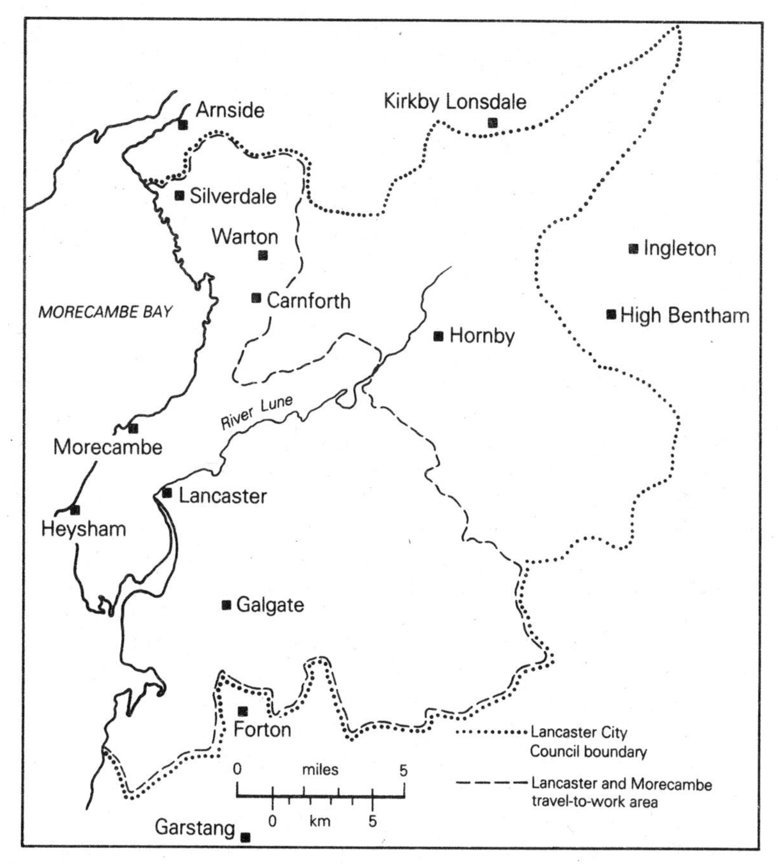

Figure 7.1 Lancaster employment and administration areas.

'de-industrialization' of the local economy occurred earlier and more quickly than in most other areas. This was not a function of the sectoral balance, which was very favourable, but rather that the existing firms, which were among the market leaders in the 1940s and 1950s, failed to respond to the changed conditions of the 1960s. Williamsons, for example, the best-known lino manufacturer, was forced into a merger with another company, and then, in 1967, made large redundancies. The other major firm dating from the nineteenth century, Storeys, which had diversified into various plastic-coated products, reduced employment throughout the 1970s, finally closing its main site in 1982. The leading artificial fibres company, Lansils, suffered job losses during the 1970s and finally closed in 1980. And, at the same time, the area failed to attract any new substantial manufacturing plants, even during the heyday of British regional policy from the later 1960s to the early 1970s. Lancaster acquired Intermediate Area status, but never had the political strength to gain higher status.

The result of the decline in the existing manufacturing industries and the failure to attract new ones was a 47 per cent reduction in manufacturing employment between 1961 and 1981, and a reduction of nearly 30 per cent between 1981 and 1984. At the same time there was growth in the public services of health and education, but the working-class holiday trade in Morecambe suffered dramatic decline. Between 1973 and 1987 the number of small hotels and boarding houses fell from 640 to 247, and the number of serviced bed spaces fell from 12 340 to 7115 (Bagguley *et al.*, 1990, chapter 3).

The pattern of industry and employment has thus been transformed during the last 25 years. All the older industries – floorcoverings, wallcoverings, plastics, artificial fibres, clothing, footwear, and resort tourism – have declined considerably or even disappeared. Some new areas of employment have developed, especially in public sector services and in the energy field (two nuclear power stations and a support base for the Morecambe Bay Gas Field).

In the rest of this chapter I shall consider some of the economic and planning policies developed locally in relation to these dramatic changes. Two particular strategies will be described and evaluated: that of encouraging 'small firm manufacturing' and that of 'tourism'. The first strategy will be discussed in sections 2 and 3. It will be shown that policies concerned with encouraging small manufacturing firms emerged fairly early in Lancaster and, indeed, that a relatively wide range of policies has been pursued, even though local government has been Conservative. Some assessment will be provided both of comparative spending on such policies and of their consequences for new firm formation rates and employment in manufacturing. It will be noted that the most important determinants of the condition of the Lancaster economy have been the patterns of expansion and contraction of various state-related areas of employment, something which local councils have more or less no ability to influence except in relationship to the original decision to site them in the area. In Section 4, I shall consider more briefly the strategy of encouraging tourism. In this case, issues concerned with the built environment are particularly important. Buildings in some sense epitomize the 'locality' and different individuals and groups seek either to preserve or to change them. I shall thus conclude with some discussion of the relationship between policy and place, where place is understood in terms of its very buildings. I shall consider how the inherited built environment in part structures the possibilities of devising effective local policies within particular places, especially where vested interests are able to mobilize around the protection or the transformation of that environment.

2 Economic initiatives towards manufacturing

Two main developments affected the Lancaster economy in the early 1960s. The first was the attraction of the new university, which admitted its first students in 1964 (see McClintock, 1974, chapter 1). Symbolically, the university's initial building had been a factory owned by Gillows, who from the eighteenth century onwards made high-quality furniture, especially for ships. The council, and particularly the town clerk, played a major role in ensuring that the university was sited in Lancaster. When it became clear that the University Grants Committee was going to insist on a 'greenfield' site (this was the era of rampant 'greenfieldism'), 200 acres of farmland were secured almost overnight and the boundaries of the city were redrawn to ensure that this land came within them. The site was presented to the university for an annual rent of £1. The council even acquired farming land elsewhere so that the tenant farmer on the site could be relocated. In many ways this was the most dramatic and successful economic intervention in the area (achieved at the expense of many other places, including Swindon and Cheltenham) and has had a long-term impact on the local economic and social structure.

The second main development was the establishment of the first industrial park in the area, at White Lund, on land that had been designated for an airport. A joint committee was established in 1964 between Lancaster and Morecambe councils, who each owned parts of the land involved. The two councils divided up the various tasks involved in running such an estate and the first factories were established during 1965–6.

Within a year or so of both these developments, further proposals were initiated, partly because of major redundancies at Williamsons. In 1967 an action group was formed to fight for government help for the area. This group consisted of representatives of the local councils, industry, trade and commerce, and trade unionists. In May 1967 the 'Action for Industry Campaign' began. In June 1967 a more specific plan was announced for collaboration between Lancaster and the university. The idea was to attract small, science-based industries by providing them with factory space, research facilities, housing for key workers, and so on. There was involvement of the university, the Chamber of Commerce, the town clerk and various other officers. The Trades Council complained that it was not properly represented. A central role in this development was again played by the town clerk,

Don Waddell, and the first vice-chancellor of the university, Charles Carter, who was also chair of the North West Planning Council, which had specifically recommended that the Lancaster area should expand employment in new science-based industry and in office employment. Thus there was strong professional support for the idea of developing an agency that would use the skills available at the university. There was little political conflict about this proposal, partly because of the relative lack of information at the time on such matters available to the elected councillors, and because there was a widespread feeling that 'something' should be done to help the area.

One further point should be noted. In letters to the local newspapers, extensive reference was made to the strengthening of regional policy by the Labour Government nationally. It was believed that such developments would seriously weaken places such as Lancaster that at the time were ineligible for such grants. This was argued for two reasons: first, that *new* footloose firms would not be attracted to the Lancaster sites; secondly, that existing plants in the area that were owned by large organizations based elsewhere would be likely to concentrate further developments at their other sites in development areas. Thus, local employers actively turned to the local authority for improved assistance to industry, partly because of the perceived threat from nationally organized regional policy, which would most benefit Labour-voting industrial cities and regions. A further factor that reinforced this feeling locally that something 'should be done' about the contemporary level of unemployment was the election in 1966 of Lancaster's one and only Labour MP, Stan Henig, who was at the time a politics lecturer at the university. Some influence was exercised in the evolution of an interventionist policy locally, by members of the professions who had no contact with the existing older manufacturing interests in the city.

Thus, in the period 1964–7 a number of influences coalesced to generate an economic project for Lancaster (Morecambe was at the time a separate council and did not develop any such project). This project involved the development of small manufacturing firms, ideally in new science-based industries. It was very much an officer-led initiative, in which interests associated with the newly arriving university played an important role. Clearly, the redundancies being declared locally provided an important context for such policies, and enabled its proponents to convey the view that something needed to be done very quickly to counteract the potentially rapid decline in manufacturing employment. It was also important that existing manufacturing interests would not feel that such a policy would threaten their interests, because with unemployment rising above the national figure there was unlikely to be any general shortage of labour in the future. It is important to note that what was established was an agency, Enterprise Lancaster, which possessed a degree of operational independence, although it was funded by the City Council. The reason for establishing this kind of body seems to have been the belief that it could operate more independently and less bureaucratically than had been the case with previous council industrial initiatives.

When the first head of the agency started work he had no models upon which to base this work. There were almost no other agencies in Britain at that time specifically concerned with local economic development. He was aware of only one other agency operating in that period and notes that, as a result, Enterprise Lancaster 'as a scheme . . . created a lot of interest . . . we had people coming from all over the country, and abroad'. At the same time he was given little direction by the Council as to how he should proceed:

> I'd no real terms of reference at all, so one more or less had to start from scratch and do a fair bit of basics . . . to find out what really existed . . . the Chamber of Commerce was dead, there was no, or virtually no information about what was going on, nothing in the town hall. (Interview, D. Kelsall, 1986)

He thought that particular attention should be paid to developing new companies and that the best way of doing this was by providing a 'seed-bed' location. This came to be located in the initial home of the university, St Leonard's House, which gradually became vacant as the university moved to its new greenfield site. The council was thus able to provide good quality initial accommodation to newly starting firms. Rents were fixed at a fairly low level, around cost, and easy entry and exit terms were arranged. It was intended to attract companies involved in research and development, design and high technology. Other attractive aspects of St Leonard's House were that the building was very conveniently located near the city centre and that a mixture of offices, small workshops and laboratories were all provided under one roof. There was a modest provision of common support services, including secretarial services for the start-up period.

Several of the companies that started in this seed-bed site have been assisted to move out into larger premises on the various estates in the locality, owned by private companies, the City Council, Lancashire Enterprises Ltd (LEL; see below) and English Industrial Estates. The city-owned estates have been developed in different periods with finance under the 2p rate; from the EEC under the Textiles Programme; and from central government under the Derelict Land Act.

Enterprise Lancaster also organizes a variety of financial inducements to both new and existing companies. Between 1981 and 1986,

56 firms received a total of 78 grants, loans and rent concessions, involving almost £600 000 (*The Visitor*, 21 January 1987, 11). At the time of their application these firms employed 91 people; they now employ 248, and another 210 are employed by two other companies that have been assisted recently. The provision of financial assistance to these companies partly results from the widely held belief that encouraging indigenous small firms, especially in manufacturing, is highly desirable. It is thought that large, externally controlled firms have caused major problems for the local economy. In 1977, it was noted that 'it is significant that a very high proportion of closures and redundancies declared during recent periods of recession have been in firms who are under external control, i.e. "pruning the branches to encourage growth" ' (Lancaster City Council, 1977, Appendix IIIc).

Enterprise Lancaster has also been concerned with more general promotional work for the locality. The agency has three main advantages and three main disadvantages in its efforts to construct Lancaster as a place in which to invest and in which skilled labour would like to live. Its advantages are, first, its generally pleasant environment. As *The Architects Journal* states:

> Lancaster is a city of great character and small size, set in some of the most beautiful country in Britain, and so the pleasures of exploring it are great (quoted in *Enterprise Lancaster* promotional material).

A rapidly expanding firm locally is Reebok UK (Fleetfoot) Ltd; one of its founders, the athlete Chris Brasher, explained that they set up in Lancaster because: 'We found in other towns the planners had ruined them. Here, the whole environment was superb. One could live by the beach, on the river, in the city or out in the Pennines or the Lake District' (*Lancaster Guardian*, 10 July 1987). In some more recent promotional material this message is rather less developed and more emphasis is placed on 'Lancaster – A Centre for New Industries and Technology', with current investment in the offshore and nuclear industries in the area worth £4000m. The second major advantage is that the labour force can be presented as relatively adaptable and unlikely to engage in industrial action. The fact that the local strike rate is about 30 per cent of the national average is featured in some of the promotional material. The third advantage is Lancaster's location, adjacent to both the motorway network and the main London–Glasgow railway. The proximity to the motorway network is reflected in the name of one of the recently established industrial estates, Junction 34.

There are three main disadvantages of the Lancaster area in terms of how it can be presented. First, it has never possessed Development Area status and, in 1984 it lost Intermediate Area status. It was therefore unlikely to be able to attract really large manufacturing plants, although they are energetically sought. Secondly, there has never been an abundance of skilled labour. In 1981, the proportion of skilled manual workers was 15.6 per cent compared with 18.5 per cent in Lancashire as a whole. Thirdly, Lancaster is geographically and socially distant from the main centres of economic and political activity in Britain. Therefore, Enterprise Lancaster has had to work hard in order to bring the place to the attention of potential investors. The main thrust of economic activity has been to help existing small firms to grow. The current Industrial Coordinator notes that, there are 'probably more long-term benefits through helping local companies set up and getting started in the local area'.

In addition to Enterprise Lancaster there are two other important agencies operating locally. The first is the enterprise agency, Business for Lancaster, formed in 1982 under the auspices of the organization, Business in the Community. Business for Lancaster provides free and confidential advice to both actual and potential businesses. In three and a half years they dealt with 950 clients and, of these, 170 or so involved new business starts. Each new business generates an average of 1.5 jobs, mainly in the Lancaster rather than Morecambe part of the locality.

The development of agencies such as Business for Lancaster represents another innovation designed to bypass established processes of democracy at the local level (see Dearlove, 1985, pp. 8–9). They are established with a director and an executive committee chosen from the sponsoring organizations. The *Guide* provided for running such agencies states that they 'should be dependent on private funding and, while local authorities may be partners, should be private sector led initiatives' (Department of the Environment, 1982, p. 8). However, the Lancaster agency is much more a collaborative exercise between the Council and the private sector.

Lancashire Enterprise Ltd (LEL) was established on a different basis. It was set up by the Labour-run Lancashire County Council in 1982 as one of the first generation of Enterprise Boards, which are often to be seen as part of the development of a 'strategy for labour' (Bennington, 1986). The current managing director of LEL argued:

> I don't think there's much evidence that really shows that in a genuine partnership sense, you get the public and the private sector working together. You get the public sector subsidising elements of what the private sector's going to do anyway, but you don't to the same degree get real partnership schemes and a lot of what we are doing now is to package up schemes where there is private money, European money, local government money, all working to a common end. So I think what we are doing is quite distinctive in that sense. (Interview, David Taylor, 1986)

Although there are parallels between LEL and the enterprise boards set up by the metropolitan counties, there are some interesting differences. The other boards were principally concerned with corporate investment and what was perceived to be the equity gap within certain industries. The approach of LEL, by contrast, has been directed both to a wider variety of industrial sectors and a more diverse set of initiatives, including training, investment, property acquisition and development, cooperatives and urban regeneration. Another point of difference is that the earlier enterprise boards sought to influence employment practices among the companies in which they were intending to invest. In LEL's case, it is merely made clear that there are various kinds of legislation relating to employment practices and that it is expected that these will be adhered to. LEL also has not spent enormous amounts of time or devoted great resources to planning elaborate strategies, in the way that the Greater London Enterprise Board did. By contrast, the approach has been in part reactive. Two main principles seem to have been in operation: first, LEL should be active in all fourteen Districts within Lancashire; secondly, no particular industrial sector should be especially favoured. The approach is described as one involving 'common-sense' rather than a grand plan. This does cause some problems because there is a dilemma about whether investment should occur in areas of greatest need or those of greatest opportunity.

LEL was originally viewed with considerable suspicion by existing organizations, such as employers groups, Conservative district councils and the various enterprise trusts/agencies. However, because its investment record appears to have been fairly successful it now has much better relationships with such groups, for example, with the Lancaster Chamber of Commerce (Hetherington, 1987). LEL had expected that they would get on better with Labour than Conservative councils, but this has not turned out to be the case. Three authorities with which they have good relations are Blackburn (Labour), Rossendale (Conservative) and Lancaster (Conservative). LEL also has good relations with Enterprise Lancaster. It was suggested in interview that this very positive response in Lancaster partly resulted from the 'entrepreneurial' zeal of the town clerk and the city architect. A city councillor maintained that: 'those who are on the City Council . . . say that the councillors do nothing anyway and say that the Council is run by the officers' (Bagguley and Shapiro, 1986, p. 34). Partly because of this LEL's premier site is in Lancaster, at Whitecross, the former industrial complex owned by Storeys. Whitecross illustrates the exceptionally wide variety of LEL's activity: it includes property development, corporate investment, industrial redevelopment, tourism, training, leisure expansion, housing, cooperative development, etc. By late 1987 there were 55 users, employing about 420 people. LEL is thus a rapidly expanding economic development company which is increasingly moving into urban regeneration (Hetherington, 1987; see also Urry, 1987b, more generally).

I have so far set out the range of local economic initiatives principally directed towards the encouragement of small manufacturing firms. Overall there appears to have been a fair degree of activity locally, particularly encouraged by the city council permanent officers. I shall now consider how Lancaster's spending compares with that found in other authorities.

First, survey data by the Chartered Institute of Public Finance and Accountancy (CIPFA) for 1984–5 shows that in terms of expenditure on 'economic promotion and development' the Districts in Lancashire fall into three broad groupings (bearing in mind discrepancies in the ways that different authorities complete CIPFA statistics). High-spending councils include Blackburn and Burnley, medium-spending councils include Blackpool and Lancaster, and low-spending councils include Preston and Rossendale (see Armstrong and Fildes, 1988). Lancaster had the fourth highest capital expenditure of the 14 authorities, and the seventh highest revenue expenditure. Compared with the other localities reported in this book, Knowsley, Middlesbrough and Thamesdown had very high expenditure, the Isle of Thanet and Lancaster medium expenditure, and Cheltenham low expenditure. Lancaster remained in the medium expenditure category throughout the 1980s.

Also found in Lancaster was a fairly strong commitment to a *wide* variety of policies, although the amount spent on each has been fairly modest. This resulted from the city council's policy of meeting the expenditure targets set by central government, of ensuring that the level of rate-borne expenditure remained below the rate of inflation, and that the rates should stay more or less constant. Nevertheless, Lancaster did provide grants and loans, something which was quite unusual in the early 1980s. At that period, 72.7 per cent of District Councils did not make grants, 88 per cent did not make loans, and 93 per cent did not make loan guarantees (Armstrong and Fildes, 1988, Table 11). In a 1984 survey less than half the responding authorities had provided 'financial assistance' in the period since 1980. Likewise, only 70 per cent of all councils are involved in general promotional work for their area. Lancaster not only does this but also provides a trade directory, a service that is found in less than half the respondents to the 1984 survey (Mills and Young, 1986, pp. 103, 108). The main limit on activity locally is now the lack of industrial land; only 45 acres were available in 1984–5. A considerable amount of reclaimed land is proposed for industrial use, but this still awaits development. About one-third of the discretionary rate borne spending power available under section 137 of the Local Government Act is spent locally (on industrial assistance). Although this proportion may grow, and this would probably involve reductions elsewhere in expenditure, it does not provide very large amounts to finance major schemes of land reclamation, the building of premises, and the provision of loans and grants to large companies. By 1988 the lack of large sites for new developments was causing serious difficulties, and potential manufacturers were said to be looking elsewhere. I shall now consider what the effects have been of this fairly wide range of relatively modestly financed policies.

3 The impact of local policies

Assessing the impact of local authority initiatives is notoriously difficult. In the case of Lancaster, we have considered the rate at which *new* firms have been established during the 1980s by using data that records the stock, starts and stops of firms registered for the payment of value-added tax (see Bagguley *et al.*, 1989, chapter 2; Ganguly, 1984, 1985; *British Business*, 1986, pp. 6–7, for some of the problems associated with this data). In most sectors in Lancaster, as in the UK as a whole, the number of companies registered in 1986 was within 10 per cent of the number registered in 1980. The main exceptions were the increases in the number of units in production (which rose 36 per cent in Lancaster) and in the number of enterprises concerned with the provision of professional, financial and personal services (PFPS) (which rose 43 per cent in Lancaster Travel-to-Work Area). National figures (1980–5) are generally similar in terms of the direction of changes, but the increases were far smaller, the number of production units, for example, only increasing by 14 per cent. This substantial increase in the number of production firms locally resulted from *very high* start rates. In 1982 it was 18 per cent, in 1983 22 per cent, in 1984 14 per cent and in 1985 16 per cent, much higher than any other economic sector. Thus, it would seem that local economic policy in Lancaster may have *helped* to create a healthy climate for the establishment of new small 'production' enterprises. The rate of new firm formation has been high, although clearly it is difficult to disentangle the different factors that have been involved, including the impact of the two nuclear power stations.

The important role of the council can be seen in another way. A survey undertaken in 1986 of all Lancaster manufacturing firms showed that of the 55 firms that replied, 39 (around 70 per cent) had some contact with the city council. In many cases this was described as considerable or very considerable and has involved assistance in the forms of loans/grants/premises/advertising, etc. (Lancashire County Council Economic Appraisal Survey, 1986). This contact also results from the fact that the City Council has been fairly successful in obtaining grants for various kinds of work. For example, it obtained £1.9m of funding under the EEC Textile Programme, more than any other Lancashire area. Also under the Manpower Services Commission Community Programme it obtained a ratio of places: unemployment almost three times the national average (Lancaster City Council, 1986, p. 13).

There were, however, three limitations on this strategy of encouraging small mainly indigenous manufacturing firms, limitations that became increasingly evident during the 1980s. First, there were large job losses within the manufacturing sector in the 1980s. Notified redundancies in the Lancaster Travel-to-Work-Area between 1979 and 1985 break down as follows: primary 0.3 per cent; manufacturing 58.7 per cent; construction 30.9 per cent; services 10.1 per cent (Lancaster City Council, 1986, p. 5). The proportion of the workforce in manufacturing fell from 21 per cent to a tiny 15 per cent between 1981 and 1984, despite the large increase in the number of new firms.

Secondly, one condition under which employment growth can occur on a sustained basis is where there is a complex of firms that feed off each other in a 'technology-oriented complex'. The best-known in the UK is Cambridge, where 322 high-technology companies had been established by the end of 1984 (Keeble and Kelly, 1986, p. 81). There has been nothing like the same spin-off in the other university/polytechnic towns in the UK. In Lancaster there are some high-technology firms that were started by people connected with the university. And one-half of manufacturing firms locally claimed to have some contact with the university (Lancaster Economic Appraisal Survey, 1986), one example being the much-publicized contact between Nairns (formerly Williamsons) and the University Engineering Department (Halsall, 1988, p. 27). However, the overall effect has not been very marked.

Thirdly, by far the most important determinants of the condition of the local economy have been the patterns of expansion and contraction of the various state-related areas of employment (and their local impacts). I have already shown that the city council played a considerable role in the attraction of the university in the early 1960s. But it clearly cannot do anything about the policies of the government or of the University Grants Committee, which have meant that higher education does not provide the expanding employment opportunities that might have been predicted. The city council, and especially the town clerk, also played a major role in the attraction of the support base for the Morecambe Bay gas field. However, locally based firms have not been able to take much advantage of this development and relatively small amounts of new employment were created (*Lancashire Guardian*, 23 January 1987). The two most important areas of employment in the 1980s are the health sector and nuclear power station construction; 12.3 per cent of the local labour force were employed in the health sector in 1981. Employment, however, will fall in the next few years as the 'community care' programme will eventually result in the net loss of a thousand or so nursing posts. In 1986 the local labour force employed in building and operating the two 1.5 MW nuclear power stations was 7750, nearly all of it full-time (and 94 per cent male). The construction workforce employed on Heysham 2 was at the time the largest in the UK (Bagguley and Shapiro, 1986). Since then there has been a notable decline, because the station has been completed.

Thus, by the 1980s, major problems have emerged with the strategy of encouraging small manufacturing firms. Indeed, many of the policy-makers interviewed were clear about the impossibility of reversing most of the major trends occurring at the level of the national and international economy. The initiatives discussed here are on a relatively small scale, so the employment consequences are likely to be fairly limited – they were worth doing but few people expected them to compensate on any scale for the extraordinarily rapid processes of economic change that have swept through the UK economy in recent years. It is also the case that any particular authority (at least outside the South-East) is likely to develop such policies simply because many other comparable authorities are doing so as well.

In the next section I shall turn to a second strategy that has been pursued particularly in the 1980s, namely to construct Lancaster as a centre for tourism.

4 The tourist strategy

In the last section we saw how Lancaster, particularly as defined and interpreted by local council officers, has responded to the recent patterns of dramatic economic and social change. The area consisted of Lancaster city, Morecambe, Heysham and the surrounding rural areas. It is a relatively self-evident 'locality', constituted in terms of a local labour market, that is, a geographical area in which most people who live there work there and vice versa. Reference has also been made to Lancaster District, the boundaries of such being reasonably similar to those of the Travel-to-Work Area (less so after 1984). However, it should not be assumed that these presumed interests of the locality are coincidental with the actual interests of those who happen to work within the boundary of that locality. In this section, I shall consider the issue of just what is a locality and therefore what is a local policy. I shall do so by examining briefly 'tourism' as a strategy. Central here is a concern with *the* locality, with exactly what kind of place it is and with how different social groups seek to establish a hegemonic strategy for the area. One important feature has been an apparent spatial division of interest, between Lancaster City, which is often thought to be increasing in prosperity and in its levels of amenity and environmental pleasantness, and Morecambe, which is progressively viewed as run-down and caught in a circle of neglect and dereliction.

In the first few decades of this century 'industry' was to be found in Lancaster (City) and 'consumption' in Morecambe. They were, in effect, opposites in a single system nationally of production/ consumption. Morecambe was the quintessential holiday (and retirement) resort, providing contrived and regulated pleasure particularly for the working class of the West Riding of Yorkshire. Lancaster had a proletarianized (and male) work force. Its skyline was dominated by the extraordinary Ashton memorial, an immense baroque folly constructed by the leading employer of the time and representing Victorian and Edwardian civic paternalism. By the middle of the twentieth century both places were prospering. Morecambe especially benefited from the post-war domestic holiday boom and had acquired one of the first indoor shopping centres in Britain. However, this was soon to change, as Lancaster by contrast gained a university, a college of education, an expansion of its large National Health Service and private complement of hospitals, an Arts Council-funded theatre, a literature festival, many new shops, restaurants and small service firms, an attractive river frontage, four museums, the listing of 270 buildings, and so on. Lancaster thus became a centre for consumption as industry has departed for industrial estates beyond the immediate city centre, and for greenfield sites in the rest of the UK or abroad. So it is being reconstructed as a 'post-modern' centre for consumption, particularly one preserving the shells of past rounds of economic activity to house new functions: the old customs house as a maritime museum, the conversion of the warehouses on the waterfront into flats, canal-side mill buildings as new pubs, and so on. As Britain is de-industrialized so places that present certain images of that past have become immensely attractive objects of the tourist gaze (Urry, 1988).

There were a number of conditions that facilitated this transformation: deindustrialization, which had made available a wide range of empty Victorian buildings; the growth of public sector employment, which had helped to generate a strong conservation movement, to the extent that both the current multi-storey car parks are clad in stone;

cultural changes in the wider society, which had produced a much larger market for 'heritage tourism', not only of grand buildings from the past but also of the vernacular buildings of the mass of the population; and the fact that few large modernist developments had been built in the city centre in the 1950s and 1960s (much less so than in Morecambe).

More recently, a Lancaster Plan has been developed that will, among other objectives, increase the shopping centre by nearly 170 000 sq. ft. The Plan preserves most of the heritage resources and puts them to work in generating a more extensive site for tourism, leisure and consumption. It involves the preservation of existing frontages and alleyways, the demolition of some modernist buildings, the creation of covered courtyards, galleries and fountains, and the use of glassed-over structures to link the different parts together. It might be said to result in a 'quaint' townscape, but one involving playful, eclectic and pastiched elements. It is 'post-modern' and, as such, generated intense local opposition, with more than 36 000 names on a petition objecting to the changed site for the Lancaster market, and more than 1400 written objections lodged within the planning period.

There is not space here to consider all the sources of this opposition (see Bagguley *et al.*, 1989, chapter 5). However, several points are worth noting. First, many protesters believed that the interests of 'Lancastrians' were being sold out to external concerns, particularly to large property and retailing companies. There was a widely held belief that this is 'our town' and that these outside concerns should play no part in its development. Many of the supporters of this view were not born in Lancaster but have arrived in the past decade or two. They often have service class jobs and their concern for preserving the 'character' of the town stems in part from their lack of mobility. No longer are such people to be thought of as 'spiralists' (Savage, 1987). There is also the belief that some of the proposed developments are on too large a scale and will involve styles of building that are inappropriate for the town. Many protesters were motivated by a longstanding hostility to modern architecture (although the Lancaster scheme is much more evidently post-modern). Such developments, it is thought, may harm the 'heritage' and hence damage the 'heritage tourism'. It is also believed, especially by the market traders, that the existing layout of buildings, activities and streets should be preserved. For them, the locality *is* the current layout. And, finally, it is argued, particularly by those living and working in Morecambe, that this development will attract shoppers, tourists and income away from Morecambe. They view it as yet another example of how the city council puts Lancaster's interests above those of Morecambe and of its rapidly declining 'resort tourism'. The development of retailing and tourism ('consumption') in Lancaster is thus viewed as a further way in which what was distinctive of Morecambe has been increasingly eroded, a point made forcibly by the new and successful local party, the Morecambe Bay Independents (Sharratt, 1988, p. 3).

This last point relates to what has been an important theme in local politics since local government reorganization in 1974, namely the development of a strong Lancaster–Morecambe split. Issues have in part become territorial rather than partisan. Lancaster councillors tended to argue for spending on jobs and industry, while Morecambe councillors wanted infrastructural spending on tourist-related projects in the Morecame area. A particular issue is the funding of the illuminations, which cost more than £200 000 p.a. and attract around 80 000 visitors. A Labour councillor stated:

> Lancaster Labour Party has often tried to put forward policies, say, cutting out the Illuminations and so forth, which Morecambe Labour Party people have opposed, and it's within the Tory Party as well . . . one of the big issues there is the . . . funding for the Duke's Playhouse in Lancaster which Morecambe councillors are against. (Interview, S. Henig, 1985)

This territorial conflict has been exacerbated as Morecambe's fortunes have declined; it came to a head over the expensive refurbishment of the Ashton Memorial within Lancaster's Williamson Park. Indeed the coastal area is becoming more a site for certain kinds of industry, for nuclear, gas and possibly oil, which all reduce Morecambe's appeal as a holiday centre. Its potential as a centre for consumption is extremely limited. Until recently it possessed no conservation areas in the centre. All of its theatres and cinemas have closed. It possesses no substantial indoor facilities. Its main asset is the magnificent view, but for that to be turned into a regime of pleasure for the increasingly mobile and choosy contemporary tourist a great deal else has to be provided. For example, in the recent 1988 *The Good Bed and Breakfast Guide* hardly any north-western seaside resort had an entry and most of the 700 recommended bed and breakfasts were located inland. There was no recommended accommodation in Morecambe, nor was there any to be found in the *Good Hotel Guide 1988*. Morecambe hoteliers have been intermittently criticized by city council officers for not providing good facilities, particularly 'en-suite'. This lack of facilities is particularly important because of the existence of extremely popular places to stay in the adjoining regions, of organized pleasure in Blackpool and of magnificent countryside in the Dales and the Lake District. There has been a dramatic decline in the accommodation provided in Morecambe, and that which remains has increasingly been taken up by construction workers, welfare claimants, former patients at the psychiatric hospitals released into 'community care', old people in converted nursing homes, and students. As a result, although Morecambe is dominated by the politics of tourism, its social composition and housing conditions are more like those of an 'inner city'. It might even be described as Lancaster's 'inner city'. Unemployment rates in some central Morecambe wards are more than 40 per cent.

As a result of the territorial competition between Lancaster City and Morecambe there is periodic debate about whether Morecambe should break away from Lancaster Council. A party advocating this, the Morecambe Bay Independents, has been active in the past two or three years. Local surveys show many people would support a Morecambe break away. (*Lancaster Guardian*, January–February 1988). Three other points are also worth noting. First, the tourist interest is itself poorly organized and by no means agreed about what would be desirable (except for a vague demand to spend more, while not increasing the rates). Secondly, it is widely believed that many of the Morecambe councillors are rather ineffective when arguing the case for tourism and hence the interests of Lancaster often carry the day by default. And thirdly, it is extremely difficult to imagine what can be done about a place like Morecambe at a time of local government budgetary constraint. One solution, reflected in a possible marina development, is to move the resort 'up-market', but because of both its historic image as a place of working-class pleasure and its inherited physical infrastructure (apart from its potentially magnificent Art-Deco Midlands Hotel; see Bagguley *et al.*, 1989, chapter 5), it is doubtful if this could work on a sufficient scale. After all, the competition now is so much more developed, not merely from abroad but also from many places in Britain that formerly were never considered to be potential tourist sites, such as Bradford or indeed Lancaster City itself, which have both been subject to Tourist Development Action Programmes. An alternative strategy discussed in the local press is to re-construct Morecambe as an Edwardian-cum-1920s seaside resort, with extensive areas made to look suitably pre-modern and probably containing at least one museum of the seaside. Both strategies would involve considerable levels of control by the council over the character of the built environment, in the former case to remove the cheap arcades and shops and in the latter case to eliminate modern shopfronts and signs and to effect a thoroughgoing 'quaintification' of the centre (Relph, 1987).

5 Conclusion

Two alternative strategies for the Lancaster area have thus been examined: that of encouraging small manufacturing firms, which was followed from the late 1960s to the early 1980s, and that of developing Lancaster's tourism in the 1980s. Neither policy has been successful in producing a rapid reduction of unemployment. Both have had their successes and failures. There has been a marked increase in the number of small firms, and yet the company upon which the council had pinned considerable hopes, which manufactured the African all-purpose vehicle for Third World countries, collapsed in 1988 among considerable recrimination. Although in March 1988 Lancaster won a top architectural award – the Diploma of Merit in the European Year of the Environment – for four restoration schemes (*Lancaster Guardian*, 19 March 1988), the hopes of reviving Morecambe's tourism look increasingly bleak.

However, this case-study does demonstrate that coherent local action can be undertaken particularly where, as was necessary in Lancaster, an important role is played by council officers advocating a fair level of public intervention. Without that intervention, the prospects for Lancaster in the next decade would be dire indeed.

Acknowledgement

I am grateful for comments on this chapter from Paul Bagguley, Michael Harloe, Jane Mark Lawson, Dan Shapiro, Sylvia Walby, Alan Warde and, especially, Chris Pickvance. I am also grateful for the comments of Dick Kelsall, formerly of Enterprise Lancaster, and Charles Wilson, City Architect, Lancaster, on earlier drafts of this paper. All errors of fact and interpretation remain my responsibility. Much of the information reported here is derived from interviews with key informants, which for reasons of space have not been separately referenced.

References: Chapter 7

Armstrong, H. and Fildes, J. 1988. 'District Council industrial development initiatives and regional industrial policy,' *Progress in Planning*, **30**, 85–156.

Bagguley, P. and Shapiro, D. 1986. *Lancaster: diversity in decline*, Paper to the Institute of British Geographers Conference, Reading.

Bagguley, P., Mark Lawson, J., Shapiro, D., Urry, J., Walby, S. and Warde, A. 1990. *Restructuring. Place, class and gender.* London: Sage.

Bennington, J. 1986. 'Local economic strategies,' *Local Economy*, **1**, 7–24.

British Business. 1986. 'UK registrations and deregistrations for VAT,' *British Business*, 19 September, 6–7.

Dearlove, J. 1985. *Restructuring for capital versus restructuring for labour.* Urban and Regional Studies, Working Paper 45. Brighton: University of Sussex.

Department of the Environment. 1982. *Local Enterprise Agencies: a guide*. London: Department of the Environment.
Fothergill, S. and Gudgin, G. 1979. 'Regional employment change: a sub-regional explanation,' *Progress in Planning*, **12**, 155–220.
Fulcher, M., Rhodes, J. and Taylor, J. 1966. *The economy of the Lancaster sub-region*. Lancaster: Department of Economics, University of Lancaster.
Ganguly, P. 1984. 'Business starts and stops: regional analyses by turnover size and sector, 1980–1983,' *British Business*, 2 November, 350–3.
Ganguly, P. 1985. 'Business starts and stops: UK county analysis, 1980–83,' *British Business*, 18 January, 106–7.
Halsall, M. 1988. 'Nairns keyed up for a revolution in wallpapering,' *The Guardian*, 1 February 1988.
Hetherington, P. 1987. 'Only winners will do . . .', *The Guardian*, 26 November 1987.
Keeble, D. and Kelly, T. 1986. 'New firms and high-technology industry in the United Kingdom: the case of computer electronics.' In D. Keeble and E. Weaver (eds), *New firms and regional development in Europe*. London: Croom Helm.
Lancashire County Council. 1986. *Economic appraisal survey*. Preston: Economic Intelligence Unit.
Lancaster City Council. 1977. *Industrial strategy for Lancaster*. Lancaster: Lancaster City Council.
Lancaster City Council. 1986. *Lancaster District: a summary of recent employment trends and prospects*. Lancaster: Lancaster City Council.
McClintock, M. E. 1974. *The quest for innovation*. Lancaster: University of Lancaster.
Mills, L. and Young, K. 1986. 'Local authorities and economic development. A preliminary analysis.' In V. Hauser (ed.) *Critical issues in urban economic development*. Oxford: Clarendon Press.
Relph, R. (1987). *The modern urban landscape*. London: Croom Helm.
Savage, M. 1987. 'The missing link? The relationship between spatial mobility and social mobility' (mimeo). Brighton: University of Sussex.
Sharratt, T. 1988. 'Neglected Morecambe bays for independence,' *The Guardian*, 30 August 1988.
Urry, J. 1987a. *Economic planning and policy in the Lancaster district*. Lancaster Regionalism Group Working Paper no. 21. Lancaster: Lancaster Regionalism Group.
Urry, J. 1987b. *Holidaymaking, cultural change and the seaside*. Lancaster Regionalism Group Working Paper no. 22. Lancaster: Lancaster Regionalism Group.
Urry, J. 1988. 'Holidaymaking and contemporary cultural change,' *Theory, Culture and Society*, **5**, 35–56.

8

Council economic intervention and political conflict in a declining resort: the Isle of Thanet

C. G. PICKVANCE

The Isle of Thanet, in the south-east corner of England, consists of the three towns Margate, Broadstairs and Ramsgate, and a rural hinterland. Its reputation is as a resort area and an area for retirement. The aim of this chapter is to outline the continuing local government role in the local economy, and the controversy it has led to in recent years owing to its relative lack of success and to the divergent interests of the local population. It will be argued that these experiences have been strongly influenced by Thanet's unfavourable position in relation to central government policy and to market processes.[1]

1 Economic development

The development of Thanet since 1900 can be conveniently divided into three broad phases: up to 1950, 1950–70, and after 1970.[2]

Up to 1950 the local economy was predominantly tourist-based. The Thanet resorts experienced their greatest success in the late nineteenth and early twentieth centuries with the rising popularity of seaside holidays and the spread of the railway network. The Thanet resorts always aimed at the lower part of the market, rather than trying to compete with higher-status seaside towns. Margate had and has the most working-class image and Broadstairs a more select image. In 1938, of the 14 000 rated properties in Margate, 6540 were used for accommodating tourists at some part of the year. Even by the late 1940s the percentage of the workforce in manufacturing in Thanet was little more than 10 per cent. After 1950, Thanet tourism went into decline as car ownership revolutionized travel opportunities, real incomes rose (leading eventually to the cheap foreign package-holiday), and cultural preferences regarding UK holidays changed, to the detriment of seaside resorts (Urry, 1988).

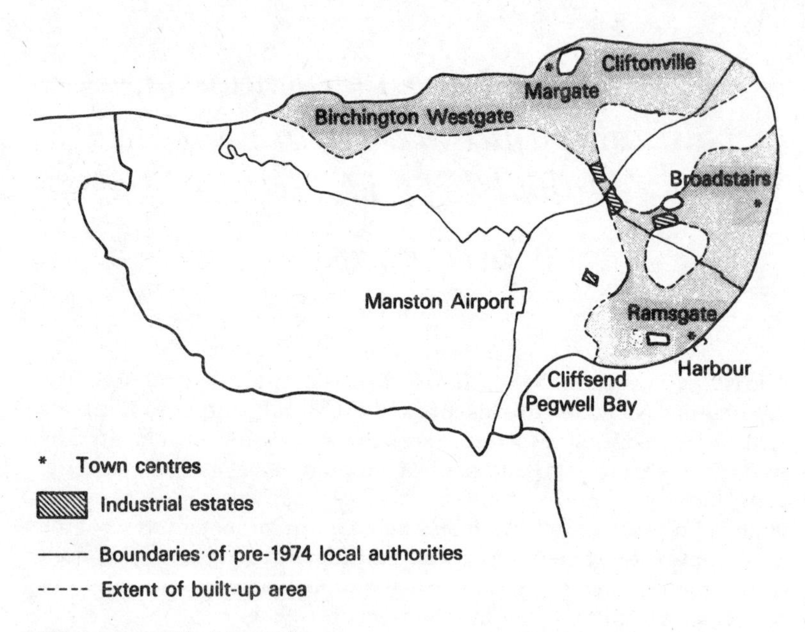

Figure 8.1 The Isle of Thanet.

By the late 1950s, the decline of tourism was serious enough to provoke council initiatives to try to attract manufacturing firms as an alternative employment source. The main initiative was a programme of industrial estate building. This added to Thanet's original attraction: a labour force used to low wages and to working in a non-union environment. A further asset was that Thanet was successful in gaining assisted area status between 1958 and 1961, and continued to receive special treatment in the allocation of Industrial Development Certificates after that. Taken together, these features helped Thanet to attract some of the many 'mobile' firms that were seeking to make complete moves from large cities or to set up plants outside them. The result was that, by the early 1970s, manufacturing accounted for nearly 30 per cent of employment. The industries involved were very diverse and the jobs were mostly in unskilled and semi-skilled assembly work. But, from the late 1970s, Thanet manufacturing experienced a sharp decline. Between 1978 and 1983, 26 per cent of manufacturing jobs were lost, and Thanet's service sector also lost 12 per cent of its jobs. This was one reason why, by 1987, the unemployment rate in Thanet was 21 per cent, the highest rate in any non-assisted area and higher than in many assisted areas; the other reason was that Thanet has a chronic unemployment problem: even without recent job loss its unemployment level would be above average.

From 1970 a new phase can be identified, characterized by the lack of any single dominant industry. The decline of tourism continued. Thanet visitors were increasingly elderly, and few new visitors were being attracted to the area. In the 1960s the number of hotel rooms had fallen by nearly half. The result today is a rump of hotels, mostly aiming at the cheaper end of the market and increasingly finding new custom among those on supplementary benefit and former patients of residential establishments closed down under the policy of 'community care'. It has been estimated that only about 20 per cent of local employment depends directly or indirectly on tourism (Buck *et al.*, 1989). There were several new developments in the period after 1970. The municipally promoted port expansion at Ramsgate started in the mid 1970s and, since 1979, a ferry service has operated to Dunkirk. There has been continued development of housing, particularly for retirement migrants. The percentage of the population of retirement age increased rapidly in the 1950s and 1960s and by 1981 had reached 28 per cent, compared with 18 per cent nationally. In this period a new services sector industry – private nursing homes – emerged, supported by a mix of public and private finance. The closure by councils and health authorities of residential facilities led to a DHSS-financed demand for hostel or nursing home accommodation, and this was supplemented by an increasing demand by retired people in need of care and able to pay for themselves. The period since 1970 is thus characterized by diversity and by repeated attempts by the council to discover or create market niches and attract new investment.

There are important continuities between these three periods that have facilitated transitions between them. First, employment in tourism, nursing homes and factories has mostly been unskilled, poorly paid, and allowed few opportunities for advancement. Much of it has been unstable: the average duration of employment in manufacturing jobs in our sample survey was one year.[3] Thanet has a shortage of higher-skilled and better-paid administrative and technical jobs, because it has few public sector jobs and lacks control functions and even routine office employment. The lack of skilled jobs has led to an out-migration of better-educated young people. For example, graduates in hotel management and catering from Thanet Technical College typically seek careers in London and abroad rather than in local hotels. Secondly, there is a local culture of individualism, enterprise and weak trade unionism, which is linked to its experience of small businesses (and above average self-employment) and the lack of large employers.

It can thus be argued that Thanet has been relatively unfavourably placed in relation to state policies and market processes. In regard to the former, it has benefited, briefly, from regional policy, and has been well placed to take advantage of DHSS-financed demand for long-term accommodation; also, recently, part of RAF Manston has been allocated for civil aviation as 'Kent International Airport'. However, Thanet has lacked the major benefits of proximity to a motorway (Lancaster, Swindon), defence contracting (Cheltenham) or the location of major government offices (Cheltenham or, in the 1960s' dispersal of government offices, Swansea, Newcastle, etc.).[4] In regard to market processes it had a strong position in the 1960s in relation to manufacturing, but failed to gain any of the insurance and other private sector offices decentralized from London at that time. Today, it has a fair position in the fields of retirement and cross-channel travel. But it is weakly placed in rapidly growing services sectors such as financial services, in office location, in retailing, in tourism and in manufacturing, where its lack of graduate labour discourages advanced firms, and its peripheral location and lack of distinctive assets, such as proximity to Heathrow or a top-rate education system, make it unattractive even to plants needing unskilled labour. The result has been to place a major constraint on the council's economic intervention, whose origins are now discussed.

2 Municipal intervention: seaside resort origins

Seaside resorts are atypical of urban areas as a whole, but have traditionally had a high degree of municipal activity. A 1911 survey of municipal spending shows that seaside resorts were uniformly in the upper half of the spending distribution and most (including the three Thanet resorts) were in the first quartile (Preston, 1985). The same pattern continues today, as will be shown below.

The reasons for the extent of municipal activity in resorts are not hard to find. While coasts and beaches are natural features, seaside resorts are social creations. Hotels and other forms of accommodation are the main requirement of a resort, but owners of accommodation are incapable themselves of providing the facilities of a resort or of promoting it – these must be collective endeavours.[5] However, the need for collective action has not automatically meant council intervention. Historical studies show that in towns where landownership was concentrated in the hands of an aristocratic or similar family, it could exercise great control over building layouts and standards, the provision of utilities and amenities, and the maintenance of social order (Walton, 1983). However, this was an unusual situation. More often, land ownership was fragmented and the council open to diverse interests. Although the evident need in resorts for a proper water supply and sewerage, for unpolluted sea water, for electricity and trams might seem to have ensured municipal action, in fact progress was very mixed owing to council reluctance to interfere with private property rights or to impose too large a burden on ratepayers. Only by 1900 had a competitive atmosphere developed among resorts and this strengthened the hand of tourist interests vis-à-vis local councils and encouraged municipal action. A 'virtuous circle', by which municipal spending could be justified because it led to a growth in rateable value, was acknowledged to exist. (Political splits nevertheless occurred about the wisdom of spending, owing to the divisions between, on the one side, hoteliers and builders and retailers closely linked to tourism and, on the other, those less dependent on it, and to the competition provided to private entertainment and leisure facilities by municipal provision.) A tradition of extensive municipal provision and enterprise thus developed in many British resorts, which we have termed 'municipal Conservatism' to make the point that interventionism is not only associated with 'municipal Socialism' (Buck *et al.*, 1989; Boddy and Fudge, 1984).

The Thanet towns are typical of other resorts in having a history of municipal action in support of tourism. This started in the late nineteenth century, before which time most theatres, concert halls and other venues, piers, foreshore activities, etc., in Thanet were provided by private enterprise. By 1906 publicity was still provided by local Chambers of Commerce on a somewhat irregular basis. From around 1900, however, there was a series of initiatives by the local councils in Thanet such as the creation of parks, the building of the Winter Gardens, the setting up of bands and orchestras, and other entertainments that supplemented private provision (Stafford and Yates, 1985). Social control functions, such as the prohibition of mixed bathing until the First World War, were necessarily municipal responsibilities. This tradition of municipal action continues to the present day. The council provides facilities that include entertainment venues in each of the three towns, leisure facilities and activities, parks, promenades, bandstands and museums. It has to meet a greater burden of spending on roads, parking and public lavatories, as well as coastal protection. And, not least, it is responsible for promoting the three resort towns.

A measure of the importance of tourism-related spending in Thanet and resorts generally can be seen in Table 8.1. This shows a significant 'excess' compared with the English average, first in total spending, and secondly in particular resort-related fields such as subsidies to trading services (e.g. entertainment facilities), parks, leisure facilities, highways and environmental health.[6]

Table 8.1 *Net revenue spending[1] expected in 1986/7 expressed in £/head*

	Thanet district[2] council	Nine resort[3] authorities	All English non-metropolitan district councils
Total net revenue spending, including debt charges	85.69	96.18	65.78
of which:			
rate fund contribution to trading services[4]	9.63	3.24	0.12
parks and open spaces	9.15	11.39	5.63
pools, sports and recreational facilities	6.52	2.47	4.55
highways and local transport	3.22	0.17	0.45
environmental health	7.69	10.35	6.16
'other services'[5]	9.08	18.79	7.87

Source: CIPFA (1988).

Notes:

1 As these figures are residuals, reflecting the balance between income and gross expenditure, they are subject to some instability from year to year.

2 The population of Thanet was estimated at 123 600 in mid-1986. An expenditure of £10 per head thus amounts to £1 236 000.

3 These are Blackpool, Fylde, Torbay, Bournemouth, Eastbourne, Brighton, Hastings, Southend and Great Yarmouth, the successors of the nine county borough authorities defined by Sharpe and Newton (1984) as seaside resorts.

4 This includes the subsidization of entertainments facilities, such as the Margate Winter Gardens.

5 This is a CIPFA category. It does not contain all services other than the five listed above, but is a residual that excludes refuse collection, town planning, non-HRA housing, rates collection, etc., on which data are provided. It is likely to contain tourism-related services not otherwise specified.

Clearly, then, the Thanet towns have a tradition of municipal intervention that is quite extensive. Paradoxically, however, the local tourist industry is one in which the main economic units, hotels, are run as small businesses.[7] The owners of hotels in Thanet are generally subscribers to an enterprise culture. Moreover, in Thanet the weak development of trade unionism means that there is considerable individualism among workers in all sectors. This tension between the two traditions of individualism/enterprise and municipal interventionism is important to the later discussion of local politics.

3 Municipal economic intervention in Thanet and its impact on rates

The aim of this section is to identify the level of council spending in support of the local economy. This is important for two reasons: to show that, in seaside resorts, economic intervention consists of a much wider range of types of support than in other types of locality; and to provide evidence on council spending as an element in the discussion of local politics in Section 5.

Municipal economic intervention is often taken to be synonymous with support for manufacturing industry. The Thanet evidence suggests a different picture. Table 8.2 shows the level of industrial support in Thanet to have varied between 0.8 and 1.6 per cent of net revenue spending from 1981 to 1986. This support consists of the provision of industrial estates, the promotion of the area to industrialists, and a small expenditure on industrial loans. But Table 8.2 also shows that, in a seaside resort, support for industry is only a small fraction of economic support, and that the total level of economic support is very considerable. *In 1986–7 close to one-quarter of all net expenditure by Thanet District Council could be regarded as support for the local economy.*[8] The main elements in this, apart from industrial support, are support for 'non-industrial' development, and support for tourism. The former involves the development of Ramsgate port, Ramsgate hoverport, and the running of pleasure harbours. Table 8.2 makes clear that in total these two types of spending dwarf support for manufacturing.

The impact of these spending patterns on rates is not direct, because council spending is partly financed by government grant. It is important, therefore, to know whether this grant compensates for the higher spending, or whether high rate levels have to be levied in Thanet to pay for this economic support.

This can be explored through Table 8.3, which shows per capita total spending, government grant and rates for Thanet and various categories of authority. It shows that, in relation to both Kent and all English district councils, Thanet spends more per capita: 37 per cent and 30 per cent respectively (line 1). But it also receives a considerably higher grant (line 2) – this reflects a central government judgement that

Table 8.2 *Net expenditure on various aspects of economic support, as a percentage of total net revenue spending (after 'financial adjustments'), Thanet District Council, 1981/2–1986/7*

	81–2	82–3	83–4	84–5	85–6	86–7
Industrial support (industrial estates, industrial promotion, industrial aid)	0.8	1.2	1.4	1.4	1.0	1.6
Support for non-industrial development (mostly port related)	0.4	10.0	4.7	7.7	6.5	3.4
Support for tourism	11.6	17.8	16.7	18.4	18.3	17.5
Total economic support	12.8	29.0	22.7	27.5	25.8	22.5

Source: Thanet District Council (1983–1987). (Where discrepancies appear in the source, the most recently published figures are used.)

Note: Due to rounding, totals are not necessarily the sum of the component figures.

Table 8.3 *Net revenue expenditure, grant and rate levels expected in 1986/7, in £ per capita*

	Thanet district council	All Kent district councils	All English non-metropolitan district councils	Nine resort* councils
1 Total net revenue spending, including debt charges	85.69	62.60	65.78	96.18
2 Expected block grant	36.97	22.69	19.68	35.94
3 Rate income	36.19	28.61	34.70	41.39

Source: CIPFA (1988).
* See Table 8.1 for definition.

its needs are above and/or its resources (in rateable value) are below the Kent and English averages. The effect of this grant is to lower the required rate income, but it remains 26 per cent above the Kent average and 4 per cent above the national average (line 3). The former comparison reflects (a) the heterogeneity of Kent and (b) the generally higher rates levels of resort authorities – the last column of Table 8.3 shows them to be 19 per cent above the English average.

It is difficult to judge the political impact of rates. Since central government grant does not fully compensate for the initial difference in spending levels, there is clearly a material basis for political conflict over rates levels in Thanet. However, whether conflict occurs depends not only on the absolute rate level but also on perceptual aspects (e.g. whether people are aware of the Thanet rate or only of their total rate, which consists mainly of a Kent County Council precept; and whether people make comparisons with Kent or with England) and on basic attitudes to council spending which, as will be suggested in Section 5 vary considerably among different segments of the population.

4 Council economic promotion

The success of a council's activity in support of the local economy depends on the context of state activity and market forces within which it operates. As was shown earlier, the context in which Thanet council works has been relatively unfavourable. In this section, some more detail is provided on Thanet council's own economic interventions. We shall leave until Section 6 an examination of local attempts to lobby Kent County Council and central government in order to change the environment within which the council functions.

The council's interventions can be divided into two major categories: support for tourism and port expansion.[9]

Support for tourism

The policies in this area consist of resort promotion, the provision of entertainment facilities, the attraction of new hotel investment and the designation of a 'holiday zone'.

Resort promotion is a traditional activity of resort councils. Up to 1974, Margate, Cliftonville (the main hotel district of Margate), Broadstairs and Ramsgate published their own brochures. In that year the newly created Thanet District Council produced a joint guide, but by 1976 dissatisfaction in the individual resorts led to separate guides being published once again. The period since then has seen persistent dissatisfaction among hoteliers about the level of council spending on promotional activity. But the basis of hotelier opposition also lies in the fact that Thanet tourism is in decline and the council is the only object on which hoteliers collectively can vent their discontent. The council thus plays a classic scapegoat role.

The dependence of hoteliers on the council is partly because, as small businesses, they lack the financial resources to promote the area to the extent they think necessary. But it is also because, as private firms in a competitive relationship, they lack the legitimacy to provide collective goods. An example of this occurred in January 1985, when a group of Margate hoteliers, centred in Cliftonville, offered to run the Thanet Information Bureau. This was met by cries of 'privatization' by Labour councillors, and cries of 'foul' by hoteliers outside Margate (and even some inside) who doubted the neutrality of the hoteliers who initiated the idea. The proposal was rapidly withdrawn.

But the hoteliers' dependence on the council is only part of the story. The council needs the hoteliers because they are an essential part of the council's vision of Thanet's economic future, despite their relative lack of success.

The provision of entertainment facilities is a second council role in the tourism field. The largest entertainment facility available in Thanet is the Margate Winter Gardens, and this is seen by hoteliers as essential to the attraction of tourists. The Winter Gardens does not cater solely for tourists. The 'summer season' (mid-July to mid-September) is shorter than previously, but outside this period both professional and amateur shows are put on. Amateur dramatic societies have successfully lobbied the Leisure Committee to obtain lower charges (and hence higher council subsidies). Another constraint on the financial situation of the Winter Gardens is the decline of factory and company dinners and dances, as factories have been shut or company social clubs have been closed.

The running of the Winter Gardens has always been problematic, and a six-figure deficit has been normal for many years. Until the early

1980s the council ran the summer entertainments, but as annual losses mounted, pressure increased for privatization. In 1981, attempts were made to get private promoters to take charge of the season, but no takers could be found. Until 1985, the council ran the Winter Gardens on a system that guaranteed performers fixed amounts. The council bore the risk and could thus do well if a show was a success, but bore the whole cost of any failures. A more than usually high loss in 1985 lead to the resignation of the chief entertainments officer in March 1986 and to (unsuccessful) calls for the resignation of the council's leisure committee. A new arrangement was established by which performer and Winter Gardens shared in the net profit of any show, e.g. in a 3:1 ratio. This solution appears to have reduced the level of losses, but has not quelled the dissatisfaction among non-tourist interests. In late 1988, the council decided in favour of privatizing the Winter Gardens, but it is too soon to know whether a taker has been found, or what financial consequences will follow.

A third aspect of support for tourism is the *attraction of new hotel investment*. Thanet tourism shows a long-standing trend of disinvestment as many hotels switch to a long-term residential clientele or are converted to flats. This process weakens the area's competitive position among tourist resorts: hence the significance of new hotel investment. During the last decade there have been proposals to build hotels on three prominent sites: overlooking Ramsgate harbour, on the cliff top at Margate, and on the site of the former hoverport at Pegwell Bay. These have all come to nothing, except for a 1985 proposal for the Ramsgate harbour site, which has now been built and is in operation as the Marina Resort Hotel. An important feature of this project was the deal it involved with Thanet district council. The council owned the land, but the developer, Resort Hotels, was reluctant to buy it for £100 000 as proposed by the council. Instead, the council transferred the land to the developer in exchange for 80 000 shares valued at £1.25. This proposal was initially vetoed by the Department of the Environment, but appears to have been allowed subsequently. It is an example of how a council can take shares in a public company. Loans were also provided by the English Tourist Board and Kent Economic Development Board. The dealing between the council and developer led to calls for the resignation of the council leader. The arrangement was seen as a typical case of the way in which the council bears an unduly high risk in order to attract private investment.

The final example of council policy in support of tourism is the *'holiday zone' in Cliftonville (Margate)*. Hoteliers within this area form the core of the Margate Hotel and Guest House Association, and individual streets also have their own organizations. The 'holiday zone' is a planner's designation, established in 1976, which indicates the council's predisposition to oppose conversion of tourist accommodation to other uses. It is intended to encourage holiday-related land uses and to discourage, for example, hostels for the mentally-handicapped. It is also seen as a way of reducing local complaints against late-night entertainment in hotels.

The existence of the holiday zone designation clearly reflects a convergence of interest between council and local hoteliers. However, it has been controversial, especially since the early 1980s, and certain streets have been removed from the zone. The conflict is economic in origin. As the fortunes of tourism fade, hotel owners are increasingly switching to long-stay residents, seeking conversion to flats, or selling up. Hence, the holiday zone designation acts as a constraint on the options of hoteliers seeking new uses for their premises. The council planners have to balance the council's desire to support tourism against the economic interests of these hoteliers. The result is that the enforcement of planning control has been made more flexible, making tourist hoteliers extremely critical of the council.

Port expansion

Apart from economic support for tourism, the council's main intervention in the local economy has been its development of the port facilities at Ramsgate. This started in the mid 1970s and involved the reclamation of land, dredging of channels and construction of a sea wall. In 1979 the Norwegian-owned Olau Line started a ferry service to Dunkirk, but this lasted only until 1980. Since 1981 the ferry service has been run by the Finnish-owned Sally Line, which subsequently became port manager as well.

The capital costs to the council of this project were large. By 1986 they totalled £7.25 million. This sum included £1.25m of capitalized interest following the collapse of the Olau ferry operation and its non-payment of rent. Moreover the revenues from the port have always fallen short of loan repayments – by £300 000 per year between 1982/3 and 1985/6.

The project has been surrounded by controversy. This has centred on three interconnected issues: the level of losses, the council's relations with the Sally Line, and plans for an expansion of the port's activities. The council is held to have been too generous in its dealings with the Sally Line, and hence to have incurred unnecessary losses. The collapse of the Olau ferry placed the council in a weak bargaining position: with a large investment in the port and with the ending of their major source of income the council was likely to view favourably any new investor. As Sally Line's own investment in the port has increased the scales have become less unequally balanced, but there is still a widespread perception that Sally has been generously treated.

In late 1984 the company put forward an expansion plan involving a new access road. This was supported by the council but opposed by many residents, who saw it as increasing the 'industrialization' of Ramsgate and who feared the increase in the carrying of hazardous substances. The costs of the road were to be shared by Kent County Council, Thanet Council and Sally Line and the access road gained a place in the county's road-building budget for 1987/8 at unprecedented speed. It remained, however, subject to a public inquiry. In June 1986 a second expansion plan was announced, involving a rail link to the ferry terminal. This also divided the local population. Under pressure from Sally Line the council gave its support. Opposition came from residents and, more surprisingly, from the council's own planning department. But the rail link plan was withdrawn in December 1986, after British Rail and the French Railways (SNCF) decided to locate the rail link in Dover. There is some debate about whether the plan was anything more than a bargaining ploy by BR. The net effect locally was to reinforce attitudes to council dealings with Sally Line, and to strengthen environmentalist objections. Meanwhile, a public inquiry opened into the proposed port access road in June 1987. Among other things, this was marked by slanderous remarks by a councillor about Sally Line, and by the appearance of the assistant planning officer as a forceful individual objector. One year later, the Minister of Transport announced his acceptance of the inspector's recommendation to reject the access road.

In sum, the major elements of council policy in support of the local economy are support for tourism and council-led port expansion. The conflicts these have led to will be discussed further in Section 5, but the main features of this intervention are that it has involved considerable subsidies to private enterprise, mostly in an indirect form such as the provision of tourism-related infrastructure or the bearing of capital costs in the case of the port. These are indicative of the council's weak bargaining position vis-à-vis new private investors and its support for a continued tourist base made up of small businesses that are heavily dependent on council expenditure.

Public attitudes to council economic action

The fact that tourism is seen to be unsuccessful and that port expansion has negative environmental effects means that support for these policies is far from universal. Before exploring the political effects of these policies in Section 5, we shall present some evidence on the complex patterning of attitudes to council economic action, based on our survey of people of working age in five districts of Thanet.

The essential findings are presented in Table 8.4. First, there is strong support for council economic intervention in general and in tourism and Ramsgate harbour in particular (rows 1, 2). Secondly, there is scepticism about the helpfulness of council action in general and on tourism (rows 3, 4), although there is confidence in the value of support for the harbour (row 5). But, strangely, action in *both* fields (tourism and harbour) is seen as helpful to employment (row 6). Thirdly, there is great disagreement in attitudes to spending, although there is more support for higher spending than for less spending (row 7). It is interesting to see that cross-tabulation shows that, of those who consider council action not helpful, many think council spending is too low: 68 per cent in the case of tourism ($n = 41$), 53 per cent in the case of the harbour ($n = 19$). This suggests that many people consider council action is not helpful because it is on too small a scale. But this is compatible with either of two conclusions: that it is effective and should be increased, and that it is ineffective (because misallocated or incompetently managed) and should not be increased.

Table 8.4 *Attitudes to council economic intervention*

	yes %	in between %	no %	*n*
Desirability				
1 Council duty to tackle local economic problems	93	4	3	95
2 Council action should be encouraged on				
(a) tourism	84		16	96
(b) Ramsgate harbour	87		12	90
	helpful	sometimes helpful	not helpful	
Helpfulness				
3 Council action on local economic problems is	9	10	80	96
4 Council action on tourism is	15	16	70	89
5 Council action on Ramsgate harbour is	53	14	33	79
6 Council action improves employment prospects	yes	not significantly	no	
(a) tourism	65	14	27	85
(b) Ramsgate harbour	78	6	16	86
	too high	about right	too low	
Spending levels				
7 (a) tourism	28	16	64	62
(b) Ramsgate harbour	22	33	47	59

Source: Project survey of 98 respondents (see note 3, p. 185). The base (*n*) of the percentages is 98, less 'don't knows' and 'no responses'.

This pattern of responses supports our earlier argument. The combination of support for intervention but doubts over its effects suggests that people are aware of the difficulty of the situation in which the council is acting. However, a significant minority considers existing spending is too high, suggesting a lack of confidence in the council. This disagreement would no doubt have been sharper still if retired people had been included in the sample.

5 Local social structure, culture and politics in Thanet

The aim of this section is to trace the effects of the policies outlined in Section 4. Before examining local political processes and the extent to which they can be seen as related to the council's economic development policy, the local social structure and culture must be outlined.

The social and cultural make-up of Thanet is largely shaped by its industrial structure and by retirement migration. The largest manufacturing firm (Hornby) has 700 or more employees (depending on the season), but the typical firm, taking all industrial sectors together, is small and non-unionized. A good example is the tourist industry, where large chains have been absent. In 1981, 85 per cent of workplaces in all sectors had fewer than 20 employees, and only 56 per cent of employees worked in workplaces with 50 or more employees. A second distinctive feature is the lack of graduate managerial, professional and technical jobs. This is due to the low order of Thanet towns in the urban hierarchy. Thanet has a lack of offices with control or sub-regional management functions (only the district health authority is quartered in the area), and there is only one further or higher education institution, Thanet Technical College. In recent years, several public sector or ex-public sector administrations, such as education and gas, have closed their Thanet offices as part of rationalization measures. To a limited extent commuting to professional and managerial jobs makes good the local deficit of such jobs.

Finally, the retired population included migrants who are more representative of white-collar and skilled blue-collar occupations than are the locals. The long-standing pattern of out-migration of 15–24 year olds, particularly those with more skills or qualifications, has slowed down since the 1970s owing to the reduced opportunities elsewhere, but is still significant. On the other hand there is a pattern of in-migration among age groups above thirty. These include people intending to set up their own businesses, and others are attracted to Thanet as a retirement area. Compared with many parts of the South-East, house prices are below average and this attracts somewhat lower-income migrants.

The typicality of non-unionized small firms and the under-representation of public sector employment help make Thanet culturally individualistic, rather than collectivistic. It is an area with a well-entrenched 'enterprise culture'. We found migrants to Thanet were often attracted by this image – in our sample survey 7 out of 50 in-migrants to the area had moved in order to set up a business, mostly hotels or shops. The local culture is therefore reproduced by in-migrants as well as locals. The relative shortage of the highly educated means that the Labour input into local politics is weak, and new social movements, such as the women's movement, are relatively undeveloped.

A final cultural aspect of the area is the persistence of distinct identities for the three main Thanet towns. This is politically significant, in that council economic intervention is judged by its impact on different areas as well as its differential social impact. In particular, there is sensitivity in Ramsgate concerning the polarization of Thanet spatially, with fears that Ramsgate is destined to be the 'industrialized' area. (In fact, the industrial estates and associated roads are situated away from the centres of all three towns, and this sensitivity is linked more to the port expansion and the impact of the resulting traffic in the centre of Ramsgate.)

Political effects

The social and cultural patterns of Thanet have direct effects on political life. One important political consequence is that a large part of the population has a strong commitment to place. This is most clear in the case of hoteliers, whose livelihood is dependent upon the image and reputation of the area yet who are mostly too small to influence this image. Hence they are vulnerable to damage to the reputation of Thanet allegedly done by the mass media or by *Holiday Which?* surveys of beach pollution. But there is also a heightened commitment to place among two other groups: for small business people generally there is a financial investment over and above the purchase of a house and, for the retired, Thanet is seen as a terminus in terms of housing moves.[10] However, this commitment to place has different political consequences for the two categories. Business people have an interest in the growth of any economic activity that increases turnover. This may translate, as in the case of tourist hoteliers, to support for council investment in the provision of facilities and intervention in the local economy, even if it means higher rates. (In our interviews, hoteliers' organizations were more concerned about competition from accommodation paying domestic rates rather than commercial rates, than about the level of rates itself.) It also includes support for regulatory intervention, such as the creation of the holiday zone. For the retired, however, the level of local economic activity and the number of jobs created are of less concern than the possible impact of council action on rates levels and on the environment. The conflict between retired people and tourist interests in resorts has been noted by Grant (1977), who also discusses the high level of associational activity of the retired and their tendency to stand as Independents in local elections.

As might be guessed from what has been said earlier, Thanet is a Conservative area. It has elected one or (since 1974) two Conservative MPs ever since the war. And Thanet District Council was Conservative-controlled from its creation in 1974 until October 1984, since when no party has been in overall control. The main features of local politics have been the split Conservative party, the existence of Independents and opposition from the Alliance rather than Labour. Independent support is concentrated in the retirement areas; in Margate, the main opposition to Conservative candidates is from the Alliance; in Ramsgate, the opposition is more evenly split, with the older council estates there the main areas of Labour strength.

The intra-Conservative split and loss of Conservative control in 1984 are closely linked to council policy regarding economic development.

From 1974 onwards, the dominant faction in the Conservative group running Thanet Council was the so-called 'Old Guard'. This group was Ramsgate-based and ideologically it was interventionist, presiding over the expansion of Ramsgate port. The minority group, known as the Rebels, were mostly from Margate, although the leaders were from Broadstairs and Ramsgate. Their position was anti-interventionist and pro-privatization. It is not clear to what extent this was linked to their immediate economic interests, or to a national level trend.

The eventual loss of control by the Old Guard followed a series of scandals, some of which made the national headlines. They included the counterfeiting of money by the chair of the finance committee, and a ruse in which the leader of the council got a friend to pose as an oil sheikh interested in putting in a rival bid at a time when the Sally Line was holding off from signing a contract. This succeeded in its aim, but was considered to have brought discredit on the council and the leader resigned his position. The April 1984 elections led to the formation of a Rebel-dominated Conservative group, which immediately sought to apply a new policy including privatization of council services, asset sales, low rates and a lower profile of intervention. (The group advocated continued support for tourism, but an ending of council involvement with hoteliers.) But the Rebel period of control was short-lived. When faced with the choice between selling assets or sacking staff there were defections from the ruling Rebel group and the leader of the group was deposed. This disloyalty led to the expulsion of those concerned by the local Conservative Associations – the former Old Guard – who became known as the Unofficial Conservatives. The equal split of the Conservative group into Official and Unofficial groups – with fourteen in each – led to the Conservatives' loss of overall control of the 54-seat council. Since October 1984, no party has had overall control.

The lack of overall control has had an impact on relations between councillors and officers. Whereas before 1984 officers knew the views of the majority group and enjoyed considerable autonomy, since then they have been more guarded in their recommendations because they are not sure how councillors will line up when the council votes. However, there seem to be considerable continuities in policy and in the conflicts it has provoked before and after 1984. This can be seen from Table 8.2 and from the discussion in Section 4, which showed no evidence of a change in 1984. This suggests that the structural features of the Thanet context, such as its weak position in relation to state policy and market processes, and its local structure and culture, are more significant than the change in council control. The diversity of the population and, in particular, the split among those who support a more interventionist economic policy by the council, and those who oppose it, guarantee that conflict will surround any and every council intervention.

6 Thanet's lobbying capacity

In this final section Thanet's success in lobbying central government and Kent County Council for changes in policy is examined.

It was shown earlier that Thanet was poorly placed in relation to state policies. This is partly because Thanet fell outside the traditional types of 'deserving' locality, such as the 'inner city' or the 'depressed peripheral region'. Its unemployment was seen as either seasonal, or as due to unemployment benefit claims by pre-retirement migrants who were not really seeking work. But this begs the question of how spatial policies such as those to benefit depressed regions come into being. They are, in fact, a response to local pressure (Pickvance, 1985), and the absence of a category of spatial (or other) policy that would benefit Thanet is partly attributable to the lack of pressure by local MPs and others in the past.

The validity of this argument can be shown by the change in Thanet's treatment by central government (and Kent County Council) since 1986. There has been a major step up in the level of lobbying and a corresponding improvement in outcomes.

This change coincides with Roger Gale's election in 1983 as MP for

Thanet North. Gale is active in pressing the claims of Thanet, unlike previous MPs who have had a more national orientation. Ideologically, he is clearly closer to the party leadership than to the maverick Jonathan Aitken, who has been MP for Thanet South since 1974. Gale is concerned to attract firms to the area, but is opposed to the principle of regional policy and enterprise zones on the grounds that they shift jobs around rather than create them. However, he is aware that Thanet is competing with areas that do have the benefits attached to such designations (including access to EC regional aid). Thanet council has made two applications for assisted area status, in 1984 and 1986, following the recasting of regional policy. Neither was successful, despite the level of Thanet's unemployment which was greater than that of many assisted areas. According to the chief executive, Thanet 'now applies for everything – even if they don't meet the criteria!'. The criteria for according assisted area status are sufficiently numerous and subjective (Pickvance, 1986) for political pressure or its absence to be a significant influence. It is likely that this in part explains Thanet's failure to secure assisted area status.

There have been two main successes for Thanet. The first, announced in December 1986, was the upgrading of the Thanet Way link to the M2, at a cost of £37m. The quality of this road has been a long-standing object of local criticism, and the replacement of the 21 miles of two- or three-lane road by a dual carriageway was seen as making Thanet feel less remote from London and hence more attractive to industrialists. This decision was the result of lobbying Kent County Council and the Department of Transport.

The second success was connected with the setting up of a government inter-departmental working party on Thanet in January 1987. This was a recognition of the fact that Thanet fell through the net of policies designed for problem areas, and its establishment was itself an achievement. Roger Gale was influential in this. The working party's report in November 1987 was largely a review of action being or to be undertaken by other bodies (Thanet Working Group, 1987). However, the report may have added to the pressure on these bodies to launch projects in Thanet. For example, the European Coal and Steel Community subsequently gave Thanet 'Employment Area' status (which brings eligibility for aid to small firms) and, in November 1988, Kent County Council made Thanet the second area in Kent to receive urban improvement funds. On the other hand, the suggestion that the English Tourist Board should declare Thanet a Tourist Development Action Plan area (which improves the level of publicity expertise available to the area) has not borne fruit – Lancaster, by contrast, has obtained such a designation.

The main scheme announced at the same time as the publication of the Thanet Working Group report was the DTI's Business Improvement Service of grant aid. This is jointly financed by European and government funds, and was presented locally as a scheme specially designed for Thanet. In fact, it turned out to be a national scheme (and one particularly appropriate to Thanet given the prevalence of small businesses) but one that was launched in Thanet.[11]

The new activism has also operated at Kent County Council level. In the past, KCC has often been perceived as the council of West Kent, due in part to the ineffectiveness of Thanet and East Kent county councillors. One index of the success of the new activism is that, prior to the public enquiry into it, the access road for Port Ramsgate was rapidly promoted to a high position in the county road investment programme.

The more vigorous pattern of lobbying in the last few years is undoubtedly in part responsible for the successes just mentioned. This is not an autonomous development. The building of the Channel Tunnel has made councils and other groupings in East Kent aware of their common interest in not being marginalized by the concentration of new development in West and Mid Kent. This has led to various short-lived formations to advance the case for development in East Kent. It has also made Kent County Council more receptive to pressure from East Kent, particularly because of environmentalist opposition to development in West Kent. The Channel Tunnel has thus created a context more favourable to Thanet's arguments, and this, coinciding with the election of Roger Gale, explains the area's greater lobbying success in recent years.

It should be noted, however, that the improvement of infrastructure and availability of new schemes of loans and grants do not in themselves result in economic growth. This depends on the response to them, and it is too early to measure this.

In conclusion, it can be seen that the Thanet case is far from an illustration of an area poorly placed in terms of all state policy and market processes, with no favourable inherited resources and with no support for council interventionism. Thanet is indeed poorly placed in relation to many other areas with regard to the state policy and market processes (though this is admittedly difficult to assess), but it has inherited a labour force used to low pay and a non-union environment, which makes it attractive to employers requiring less-skilled workers, and a built environment that opens up various opportunities for small businesses such as nursing homes. Likewise, despite the divisions within the population on council economic action, there is an inherited tradition of interventionism, which in the last year or two, aided by a new MP and fears of the effect of the Channel Tunnel, has started to bear fruit.

What is difficult to assess is whether the ambivalence among the public towards council economic interventionism is due to differences in economic interest, to judgements about the competence of local politicians and local government officers, or to the scale of the economic problem the area faces (and particularly the chronic nature of unemployment in Thanet).

Some evidence was given earlier of likely differences in interest regarding rates levels and council interventionism between small businesses and the retired, on the one hand, and hoteliers and people of working age, on the other. But economic interests do not express themselves in a vacuum, and the process of political mobilization, involving prevailing ideologies, party apparatuses and institutions, must have an equal importance. While it is likely that the Thanet population has become more diverse socially during this century, this would only lead to greater diversity of public opinion if political bodies have allowed this. The evidence presented above of political fragmentation is consistent with this view.

The topic of the competence of local politicians and local government officers, including their lobbying failure until very recently, is beyond the scope of this study. However, some links between the image of Thanet and its politicians and council employees can be postulated through selective processes such as job recruitment, or the functioning of local political party apparatuses. There is some evidence that local government professionals do respond to the image of a local authority in applying for jobs (Donnison and Soto, 1980, chapter 12) and it is reasonable to suppose that the choice of political representatives reflects the pool of local talent, and the selection processes of local party apparatuses. It is in ways such as these that local economies and local cultural patterns are reproduced.

Finally, however, the magnitude of the economic problem facing Thanet council cannot be ignored. Even if there were whole-hearted public support for council interventionism, and no question about the competence of local personnel, there would be grounds for scepticism about the likely success of council action. Local level bodies may 'make a difference' but they are only one element in the situation, and cannot be expected to transform it.

Notes: Chapter 8

1 I would like to acknowledge the stimulus of my colleagues on the project, Nick Buck, Ian Gordon and Peter Taylor-Gooby. Many of the ideas in this chapter have been developed collectively as is evident from our other publications (Buck, 1987; Pickvance and Buck, 1987; Buck *et al.*, 1989). Responsibility for errors, however, remains mine.

2 For a full account of Thanet's economic development see Buck *et al.* (1989).
3 The survey involved interviews with 98 men and women between the ages of 18 and 65, selected randomly in five areas of Margate, Ramsgate and Broadstairs. The areas included two council estates, two central areas with a lot of cheap rented accommodation, and one area with white-collar as well as blue-collar workers. The aim was to over-represent respondents with experience of lower-skilled jobs.
4 It is hard to be categorical about whether Thanet is badly placed in relation to state policies compared with other localities. To be sure, we would need a tabulation by locality of data on government offices, government policies, infrastructures and contracts. Thanet is almost certainly worse placed than other localities with similar levels of unemployment, and the same is probably also true if a wider comparison is made.
5 Certain luxury hotels in isolated locations, e.g. on the Mediterranean, do provide a wide range of facilities, but arguably these are in effect a substitute for a resort.
6 Previous studies have not always come to the same conclusion. Edwards *et al.* (1976) calculated that tourism-related expenditure was 8 per cent of total local authority expenditure in the South-West. Since this total expenditure included county council spending, such as education, we would expect the figure in district councils to be bigger. However, the study by Sharpe and Newton (1984, ch. 7) of county boroughs in 1972–3 shows seaside resorts as spending less per capita in total than other types of county borough. This was because, although they spent more on parks, police and highways, they spent much less on education. This suggests that when resorts are part of county boroughs, as before 1974, higher resort-related spending is counterbalanced by lower spending on non-resort related items. After 1974, the district council data reported in the text show higher per-capita spending in resorts because this counter-balancing effect is smaller or absent owing to the different range of functions of district councils and county boroughs.
7 This is not intrinsic to tourism – hotel chains do exist, but invest mainly in cities where there is an all-the-year-round business and conference clientele – but is a result of the historical development of the Thanet resorts.
8 In money terms, total net revenue spending on the three types of economic support in 1986/7 was £2 004 994. This was made up of industrial support £141 910 (industrial estates £30 979, industrial promotion and industrial aid £110 931), support for 'non-industrial' development (mostly connected with Ramsgate port) £306 246 and support for tourism £1 556 838. In addition, capital expenditure in 1986/7 totalled £1 001 590 – made up of spending on harbours £322 536, industrial estates £328 771, and coastal protection £350 282 (Thanet District Council, 1987).
9 Its efforts in the area of industrial support will not be discussed here. Current policy is to build units of 3000 sq. ft. or less to attract small businesses.
10 The out-migration of many better-educated young people, who lack this commitment to the place, serves to increase it among those who remain.

11 The scheme offers 15 per cent investment grants of up to £15 000 to small firms with up to 25 employees, and 50 per cent grants of up to £25 000 for feasibility studies leading to new products and processes for firms employing 200 people or less, and a variety of smaller 50 per cent grants for other commercial purposes.

References: Chapter 8

Boddy, M. and Fudge, C. (eds). 1984. *Local Socialism?* London: Macmillan.

Buck, N. H. 1987. *Restructuring and conservatism in a local service economy: tourism in the Isle of Thanet.* Paper to the ISA Conference on Technology, Restructuring and Urban-Regional Development, Dubrovnik.

Buck, N., Gordon, I. R., Pickvance, C. G. and Taylor-Gooby, P. 1989. 'The Isle of Thanet: restructuring and municipal conservatism.' In P. Cooke (ed.), *Localities*. London: Unwin Hyman.

CIPFA. 1988. *Finance and general statistics 1986/7.* London: Chartered Institute of Public Finance and Accountancy.

Donnison, D. V. and Soto, P. 1980. *The Good City*. London: Heinemann.

Edwards, S. L., Jackson, B. G., Ankers, M. G. and Dennis, S. J. 1976. *Tourism in the South West Region: Methodological Report.* London: Department of the Environment.

Grant, W. 1977. *Independent local politics in England and Wales.* Farnborough: Saxon House.

Pickvance, C. G. 1985. 'Spatial policy as territorial politics: the role of spatial coalitions in the articulation of "spatial" interests and in the demand for spatial policy.' In G. Rees *et al.* (eds). *Political action and social identity: class, locality and ideology*. London: Macmillan.

Pickvance, C. G. 1986. 'Regional policy as social policy: a new direction in British regional policy.' In M. Brenton and C. Ungerson (eds). *Yearbook of social policy in Britain 1985–6.* London: Routledge and Kegan Paul.

Pickvance, C. G. and Buck, N. H. 1987. 'Municipal economic intervention in an "enterprise culture": economic restructuring, council policy and local politics in Thanet.' Paper to Sixth Urban Change and Conflict Conference, University of Kent, Canterbury. Revised version to be published in *International Journal of Urban and Regional Research.*

Preston, B. T. 1985. 'Rich town, poor town: the distribution of rate-borne spending levels in the Edwardian city system,' *Transactions of the Institute of British Geographers*, **10**, 77–94.

Sharpe, L. J. and Newton, K. 1984. *Does politics matter?* Oxford: Clarendon Press.

Stafford, F. and Yates, N. 1985. *The later Kentish seaside.* Gloucester: Alan Sutton.

Thanet District Council, 1983-1987. *Serving Thanet (Annual Reports for the years 1982/83 to 1986/87).* Margate: Thanet District Council.

Thanet Working Group. 1987. *Thanet: a strategy for growth.* London: Department of Trade and Industry.

Urry, J. 1988. 'Cultural change and contemporary holiday-making', *Theory, Culture and Society*, **5**, 35–55.

Walton, J. K. 1983. *The English seaside resort: a social history 1750–1914.* Leicester: Leicester University Press.

9

Conclusion: places and policies

JOHN URRY

1 Locality, local interests and local policy

This book has been concerned to chart the ways in which seven localities have been able to respond to the tremendous economic, social and political changes that have swept through Britain in the past couple of decades. We have described a variety of policies pursued in these localities, which range from some of the most deprived places (Liverpool, Middlesbrough) to some of the most prosperous (Cheltenham, Swindon). In each case it has been shown that the emergence, implementation and effectiveness of local policies depends upon the complex of economic, social and political conditions found within and beyond a given locality.

Nevertheless, it might be thought surprising to be discussing localities at all. Two decades or so ago it appeared that various processes were making different parts of society much more alike (see Warde, 1985). Societies were viewed as increasingly 'organized' and industrialized. Differences between places were seen as being eroded and local specificities dissolved by broad general processes, such as national systems of education, the mass media, large bureaucracies, and so on. At the same time, theories were developed that implied that what happened in particular places could be 'read off' from the general and that studying specific 'communities' was a misguided and misleading exercise. However, more recently, several changes have led to the resurgence of interest in the study of developments that appear to have heightened local differences and the symbolization of such differences. These changes include: the increased ability of large companies to subdivide their operations and locate different activities within different local labour markets; the breaking up of previously rather coherent regional economies, the competition among councils for jobs, the growth of interregional differences, and the localizing of the previous *regional* policy; the decreased tendency for voting patterns to be determined by national processes and a heightened importance of local 'neighbourhood' effects; the enduring significance of symbols of place and location in much of the mass media; the decline in the popularity of the international style of the modern movement in architecture and the resurgence of local vernacular styles; and the resurgence in some places of local politics, partly because of a revived commitment to decentralization and partly because of the very attack of local government autonomy in the 1980s (see, for general discussion, Murgatroyd *et al.*, 1985; Dickens, 1988). These processes seem to have heightened the distinctiveness of one place compared with another, although there are, of course, some homogenizing processes operating as well.

Furthermore, we saw in Chapter 1 that localities in Britain possess important powers of government. Local authorities have political legitimacy, except in a few cases; they are responsible for the implementation of services and hence can mobilize expertise, staff, land, buildings, etc.; they can levy rates and charges; and they possess much information resulting from their 'executant' role. We also saw that the efforts of post-1979 Conservative governments directly to control the levels of local government expenditure have not been especially successful, thereby suggesting that there are here important powers. More significant, though, for local government have been the attempts in the mid-1980s to undermine the very viability of such government, by eliminating one tier of administration (Metropolitan Counties such as the GLC); by weakening the powers of professional groups within local government; and by establishing new institutions which bypass democratically elected local authorities. Such structural changes have been responses by the Conservative government to the collapse of the 'dual polity' and the resulting politicization of much local government. Their effect has been seriously to weaken the capacity of local authorities to develop coherent policies, particularly with regard to economic regeneration. What has happened has been an attack on local authorities as the provider of services, in this case, in local urban regeneration. Particular emphasis is being placed instead on the private sector, assisted by central government. This is to be seen most clearly in the Urban Development Corporations in the UK – found, for example, in Merseyside and Middlesbrough.

In this book, by contrast, it has been assumed that the interests of the 'locality' are best represented by democratically elected local authorities; that private capital will not be 'representative' of local interests; and that the direct role of central government in local initiatives should be limited on grounds of both 'efficiency' and 'democracy'. However, these assumptions do need to be examined, since they rest upon a rather simple conception of the 'local' or 'locality'. We will now consider a number of difficulties and issues that arise with regard to the idea that 'local policy', as devised by the relevant local authority or authorities, is simply in the interests of a given 'locality'.

First, the notion of interests of a locality is problematic, because localities comprise a highly diverse set of social groupings, some of which will have a clear stake in the place, others much less so. It is not necessary for people's interests to be represented territorially. Nor is it necessary that the improvement of the economic base of an area will be in the interests of all or even most of the people who live there. For example, if most of the population of a place are 'spiralists', people who spiral around the country as they are promoted upwards in their corporation, their likely interest in or commitment to that locality will be very limited (although that of their home-based partners may be quite different). Furthermore, even where there is some degree of territorial identification it does not follow that this will coincide with 'locality', which in the research reported here has been interpreted in terms of travel-to-work-areas. Indeed, for most people, locality is increasingly coming to mean travel-to-shop or travel-to-engage-in-leisure-areas. In such cases, the range or boundaries of such localities will be quite different for those with access to a car than for those without, and indeed for those with access to forms of public transport (high-speed train/air) which do much to annihilate the distances between apparently separate localities.

Moreover, different social groups will have various kinds of interest in a place. Some will benefit more from expanding the employment base of a locality, others from increasing the range of shops, others from improving the leisure and entertainment facilities, others from modernizing the physical environment, others from reconstructing the 'heritage' of the place, others from making it a safer place in which to live, and so on (see discussion in Bagguley *et al.*, 1990, chapter 5). The interests of individuals and social groups are therefore heterogeneous – ranging from the 'material' to the 'cultural' and to what can be termed the 'aesthetic'. Different social groups have different stakes in a place, and their interests vary from the more obviously material (which itself varies from the straightforwardly 'economic' to that of ontological security) to the more cultural and aesthetic. Furthermore, some social groups will possess superior sets of resources and this may have the result that their conceptions of the interests of the locality become dominant, for example, that old buildings should be preserved rather than modernized. Also it should not be presumed that individuals and social groups only act in terms of a narrow sense of self-interest. There is an important range of 'altruistic' pressure groups, who may take on what they presume to be the interests of other groups or indeed the 'locality' in general. And while such groups may or may not be correct in their 'altruistic' interpretation of the locality's interests, it is almost certain that their actions will have unintended consequences for the locality.

A further issue concerns the formulation of local policies. As was shown in the studies of Lancaster and Swindon, it was professional officers rather than councillors who played the major role in devising and pushing through interventionist policies. Indeed, it appeared that officer-led policies were 'necessary' because the elected councillors did not themselves possess the vision or foresight to be able to think through what would be strategically necessary for their area. In some cases, officers have also been important in mobilizing and orchestrating the interests of groups in the locality. This is true not only of those likely to be unorganized (like tenants), but also of those who are organized (like hoteliers in Thanet or Morecambe). Far from these groups imposing their wishes upon the elected authority, a more characteristic pattern may be that they have to be forced by the relevant officers to consider whether they do possess any interests in common and how these might be realized.

This raises a more general issue mentioned in Chapter 1 – the degree to which local service delivery is professionalized and what consequences this could have on local policies, including especially local economic policies. Until recently, central government has endeavoured to ensure that services were professionalized; this was seen as necessary in order to overcome 'localism'. The process by which this occurred was often complex, involving a variety of social reformers, who contributed what Laffin (1986) calls a 'policy community', which in effect sponsored the putative profession. Laffin discusses how such a professional grouping emerged in the case of highway engineering; and how this occurred to a much more limited degree in the case of housing management. The former profession exercises considerable influence over policy. Professionalization thus places limits on the control exercised by local authorities over their staff, especially as qualifications, training and selection have often been entrusted to national bodies such as universities and polytechnics, professional institutions and central training boards.

We shall now consider briefly how this relates to the possible professionalization of 'economic development'. First, local economic development is the province of several different individuals and agencies in a local authority and hence it is difficult to form them into a cohesive group that is able to pursue a professionalization strategy of occupational closure (Larson, 1977). Although there is a putative organization of local economic development officers (EDOs), they cannot present themselves as having a particularly distinct set of technical skills and competencies. Secondly, this is reinforced by the fact that in a period of limited new economic activity they are in straightforward competition with each other in order to attract the few footloose plants and offices available. Furthermore, economic development is a local authority activity about which most councillors will believe themselves to be well-informed. They will thus be unlikely to allow it to be treated as a purely technical matter (like, for example, road engineering). And, finally, Laffin (1986, pp. 216–9) records how in the past few years there have been considerable challenges to the professionalization of well-entrenched groups from both central government and from the lay public. This has meant that those occupations concerned with economic development, which had yet to professionalize, will almost certainly not now be able to do so. As a result, it is unlikely that there will be an entrenched group of officers in an authority arguing strongly for increased spending on local economic development (by contrast with, for example, highway maintenance). Partly as a result, most councils only spend a fraction of what they have been able to under the 2p rate raised under Section 137 of the Local Government Act 1972. Moreover, other officers in an authority, such as architects or planners, may well take the lead in economic development matters, and this may make the policy initiatives fragmented and disorganized. Thus, there may be an insufficient development of professional procedures and standards, which might have ensured that examples of good practice of economic initiatives would be quickly and effectively publicized and generalized from council to council. Localism rather than professionalism will thus be dominant in British local economic policy, especially with the growth of what was termed 'central government localism' in Chapter 1.

It has thus been necessary to set out a number of points relating to the problematic nature of 'locality'. We cannot simply presume that a given effect (e.g. of increased employment in manufacturing) is necessarily in the interests of all social groups who live and work in a given area.

2 Explaining policy variation

We shall now turn to the explanation of why particular policies are found in particular places. Although our study has an advantage over many others in that we possess relatively detailed information about seven places in England in the 1980s (rather than, say, one or two), the degree to which these permit us to develop systematic comparative theses is limited. The argument here is organized in terms of 'conditions', 'resources' and 'strategies'. 'Conditions' refers to the wide variety of circumstances, external to a *given* locality, within which local policies have to be formulated and implemented. Conditions consist of circumstances that are both constraining and enabling. They can be divided into three main categories: the economic – such as the national growth or decline of particular industrial sectors; the political – such as the changing patterns of state expenditure that have (intentionally or not) spatially discriminating consequences; and the cultural – such as the changing valuation placed upon different kinds of leisure activity.

'Resources' refer to the means that are institutionally available to decision-makers within a given locality. They can be classified as follows (see Rhodes, 1988, for a related classification):

(a) *historical:* the legacy of previous rounds of economic activity, in terms of the prevalence of particular economic sectors, a given built environment, and a labour force with particular characteristics (young/old, male/female, skilled/educated/trained).

(b) *geographical:* this includes both the physical environment, especially the land available, its appeal to different aesthetic interests, and the availability of 'natural' resources; and the location of the area in relationship to others and to major routes of communication.

(c) *financial:* the size and forms of finance available to fund different strategies.

(d) *organizational:* the capacity of major institutions (public or private) to organize appropriate people, finance, skills, land, building materials, equipment, etc., in relationship to desired strategies.

(e) *legitimacy:* the ability of decision-makers to act with public support. This depends upon dominant values, such as 'public service' or the 'ability to pay'; and upon institutional arrangements, such as voting or the sovereignty of the consumer. In certain cases, a 'hegemonic project' develops, which results from the capacity of key groups to identify a development strategy that is realistic with regard to the pattern of external and internal constraints and possibilities facing that locality.

'Strategies' refer to the various sets of locally implemented policies and procedures which have as their explicit or implicit aim the raising of the level of economic activity within a given locality. Such strategies vary on a number of dimensions:

(a) *sources of finance:* whether private/public/mix of the two, and if public whether local/central/European.

(b) *form:* whether the project involves direct investment or involvement by public bodies, or the public provision of incentives to private bodies.

(c) *sphere:* whether it concentrates on the economic base of an area and, if so, whether this is manufacturing/services; or on the social base; or on the physical environment.

(d) *direction:* whether to effect growth, or to maintain a steady state, or to prevent decline.

We shall now consider briefly the nature of the different strategies that emerged within each locality. We shall attempt to highlight why different patterns developed in the various places, paying attention to the respective conditions, resources and strategies. The first two localities to be considered are Swindon and Middlesbrough (Teesside), where in the 1960s coalitions of local interests pursued modernization or growth strategies, albeit of rather different sorts.

The conditions and resources that facilitated the initial expansion of Swindon in the 1950s were as follows: its location close to London and the deliberate national policy of overspill development after 1952; the availability of large areas of cheap land; the perception locally (with no organized middle-class opposition) that the historical legacy of a 'one-industry town' was dangerous and that the encouragement of growth and diversification was desirable; and the organizational and financial skills of a small group of council officers and councillors. When further expansion was being mooted in various south-east towns in the 1960s (that is, a new condition), the fact of successful expansion in the previous decade provided the justification for new growth also to be based in Swindon. Two sets of resources aided this development: the siting of the M4 and later the use of 125 mph trains along the rail line passing through Swindon; and the development locally of a highly streamlined administrative structure with an entrepreneurial chief executive and later with the more-or-less autonomous Swindon Enterprise agency. One condition that greatly facilitated the strategy of town expansion was the shift of employment away from inner cities, especially London, to a variety of smaller towns and cities in the South-East. This partly resulted from the cramped and expensive sites available in central London locations, as well as from a general belief that free-standing towns and cities, such as Swindon, Cheltenham, Milton Keynes, Cambridge and so on, provided a more congenial environment with fewer labour problems than did inner city areas (see Fothergill and Gudgin, 1982).

By the early 1980s a re-evaluation took place of Swindon's town expansion policy, prompted by publication of the 'New Vision' document. For a number of reasons the strategy was no longer completely viable: most development land was now in private hands because the council could no longer afford to buy it; the council

(Thamesdown) was rate-capped and development would in the future have to be much more reliant on private finance (even though this would stretch the local authority's budget even further); there was no longer a public consensus supporting expansion regardless of the environmental costs; and, anyway, expansion would be increasingly based on Swindon's role as a commuter-town – it was housing-led, as Swindon is rapidly integrated into the South-East housing market.

So, in Swindon, the conditions that had sustained the viability of what we can term a 'labourist' growth coalition had evaporated by the mid-1980s. Clearly, Swindon will continue to prosper in the sense of a very buoyant housing market and considerable in-migration of new firms. However, the development will be less subject to control than in the past, and it will not be so possible for the local council to use some of the gains from the development process to finance improvements in the social infrastructure. Swindon demonstrates that localities do make a difference, and that a growth coalition based around some 'professional' council officials, councillors, and much of the informed local population, could bring about a set of public sector-led changes of a generally positive sort. In future, though, it seems that although local growth will continue this will be more through developing privately financed sub-localities where the development gains will not be used locally at all.

There are some obvious parallels with the growth coalition found within Middlesbrough (or Teesside). Here, there was also a hegemonic strategy for the area, beginning in the early 1960s, focusing less on housing and much more on the *existing* key local industries. The strategy involved the thoroughgoing modernization of Teesside and its reconstruction as a centre for economic growth. In some ways the proposed transformation of Teesside symbolized the modernization of the UK economy, a notion that swept into public debate in the late 1950s. The conditions that made possible this modernizing strategy were: the importance of the Teesside's dominant industries of steel and chemicals (and later oil) to the national economy; the perceived political importance of the area as reflected in the involvement of Lord Hailsham, and the cultural emphasis in the period upon modernization and technological regeneration (the high point of Britain's modernism!). The resources that made possible such a strategy were the legacy of steel and chemicals production and the resulting labour force skills; considerable sources of finance, especially from central government's regional policy; the availability of large areas of land and the possibility of reclaiming additional land from the sea (stimulated by the discovery of North Sea oil and the fact that conservationism had not yet emerged as a significant obstacle); and a strong consensus on 'getting things done', which united sections of capital, labour, the local authority and the local population in a hegemonic strategy.

However, by 1983 Cleveland County had the highest registered unemployment rate for any county in Great Britain. To the extent that production had been concentrated in branch plants in chemicals, oil and steel as a result of the state-sponsored restructuring of the 1960s, Teesside was particularly vulnerable when international competition led to huge cuts in capacity in the late 1970s and 1980s. A number of limitations of the Teesside strategy became clear in this period: the changing international division of labour and the resulting peripheralization of UK branch plants; growing political opposition locally to modernization and hence a breakdown in the overall legitimacy of the strategy and less confidence on the part of the ruling group in its continued viability; the effects of environmental damage in reducing the attractiveness of Teesside to potential investors (partly reflected in the oversupply of industrial land) and in limiting its possibilities of becoming a centre for service sector growth, especially retailing; and the decisions of central government with regard to the nationalized industries, especially steel.

By the 1980s a whole range of national and international changes had occurred, which had rendered the Teesside strategy inoperable. But this is not to suggest that in the 1960s this was not the most sensible strategy to pursue; or to deny that it had some considerable effects on the trajectory of local development. Local policy did make a significant difference in that it led to the continued concentration on large-scale manufacturing industry (by contrast with Swindon's explicit diversification policy). But it did not make a difference in the sense of enabling Teesside to avoid the devastating collapse of manufacturing employment, especially in chemicals and steel from 1979 onwards.

However, the collapse of manufacturing employment in Merseyside has if anything been more spectacular, a process which had particularly devastating effects upon the outer estates – Kirkby, Speke and Halewood. These were areas that had attracted substantial numbers of manufacturing plants in the late 1950s and early 1960s under the sway of nationally organized regional policy. But these were essentially branch plants and many closed or dramatically reduced their employment from the late 1970s. By the early 1980s these areas had enormous problems – on some measures they were the poorest and most deprived in Britain. The previous strategy had failed dismally. In this decade the two Labour-controlled local authorities concerned, Liverpool and Knowsley, responded to this desperate situation by devising different strategies.

Both suffered from particularly difficult conditions and poor resources. A great many local people were employed in vulnerable branch plants in industries that were experiencing exceptionally intense international competition. There were substantial reductions in the levels of both regional policy funding and of central government funding of the local authorities. By the 1980s there were two 'unhelpful' legacies among the workforce: on the one hand, of the tradition of casualism derived from Liverpool's history as a major port; on the other hand, of the limited skills appropriate to the large manufacturing plants that had recently arrived. What was apparently lacking, certainly in Knowsley, was any tradition of small-scale 'enterprise culture'. Nor were there local supply links, which would have ensured that any new initiatives had a strong multiplier effect upon the local economy. The collapse of such industry meant that, although there was plenty of land, much of it was located in an environment that was unattractive and unlikely to appeal to potential investors. Finally, no particular political party, or smaller group in Liverpool, enjoyed legitimacy and would be seen as necessarily having the authority to carry through a wide-ranging strategy of revival. Set against these points, however, was the fact that economic restructuring was clearly having such dramatic consequences that there was an exceptional need to do something to challenge and offset such processes. Everyone accepted there was a major problem.

The Urban Regeneration Strategy in Liverpool, normally seen as that advocated by Militant supporters, involved the widespread renovation and rebuilding of the city's municipal housing stock. It did not particularly seek to attract footloose firms, partly because this had brought about much of the city's problems in the 1980s, but sought to build up employment within the council itself. The council's negative attitude to various voluntary associations, and especially to cooperatives, meant that considerable local support evaporated, most obviously in the case of the 'Eldonians', the 'Black Caucus' in Vauxhall, and the 'Speke Together Action Resource'. The 'resources' of ethnicity and 'community' could not be simply ignored or obliterated in the pursuit of an apparently class-based strategy. Knowsley, by contrast, pursued less confrontational policies. One important local resource has been the high level of community politics, such as the first trade union branch for unemployed workers and many tenants associations, including the Tower Hill Unfair Rents Action Group. However, by the late 1980s, much of the running locally in the area is being made by initiatives that bypass local authorities, the Merseyside Development Corporation, the Enterprise Zone and the Freeport.

There are some parallels between Liverpool and South-West Birmingham. They are each part of a conurbation, both have depended for substantial employment upon major manufacturing plants, and both have enjoyed few resources for local regeneration. South-West Birmingham has been a heavily dominated labour market, with two major employers, Austin-Rover and Cadbury-Schweppes. Both operate within manufacturing industries that have seen huge reductions in capacity in recent years. Particularly severe have been the reductions in the UK motor vehicle industry as a result of intense foreign competition. However, the general area of the West Midlands had been prosperous up to the mid-1970s and so there was little chance politically of acquiring much support at the national level.

South-West Birmingham's resources too were not helpful: the legacy of semi-skilled workers; an unattractive environment; a lack of land for expansion; a location within the conurbation and a political inability to devise policies specific to that particular part of Birmingham; the success in the 1960s, which meant that people did not think that the area had to 'sell itself'; the longer-term consequences of a relative absence of class or locality tradition that could have provided the basis for the development of a growth coalition for the area; and, related to this, the tradition of individual household 'coping strategies', which meant that it was unlikely that strong pressure would be mounted locally to force through alternative policies. So, although this is an area with two very well-known companies, these have not really helped, especially since Austin-Rover's Longbridge plant has not provided a positive industrial image, especially in the 1970s. Thus, unlike Middlesbrough, which had a not dissimilar industrial structure, no spatial coalition developed in South-West Birmingham, and even in the West Midlands as a whole local divisions slowed down organization. Without a locality-wide economic policy various campaigns developed on other issues, particularly health, and there was a strong emphasis upon strategies to enable individual households to survive considerable levels of deprivation and hardship.

In the other three localities, Cheltenham, Lancaster and Thanet, an attractive physical environment is of greater importance to the strategy pursued, because all are partially tourist sites and contain buildings and/ or a physical landscape that is a major resource. By contrast, the three places that we have just analysed, Teesside, Knowsley and South-West Birmingham, are among the very few localities in the UK without a tourist industry!

The conditions affecting Cheltenham have been generally helpful: unlike Teesside, it developed a strong representation in growing service industries; it has benefited from the continuing large central government expenditure on defence-related industries (including that on GCHQ); and it has gained from the shifts in patterns of contemporary tourism, which have favoured inland places with well-preserved old buildings. It also possesses fairly good resources: its historical legacy was of a diversified economy, little heavy industry and the aesthetically pleasing Cotswolds and the inherited built environment; it is situated in the southern half of the country and has been an admirable site (like Swindon) for the relocation of offices from the 1960s onwards; it has a political conservatism which, even in the later 1960s, had the effect of producing considerable support for conserving the environment, especially the Regency facades, at a time when elsewhere the modern movement in architecture was running rampant (as in Swindon, Birmingham, Liverpool and Teesside).

There was, nevertheless, considerable political conflict regarding a strict conservation policy that was finally adopted in 1974. What has turned out to be in Cheltenham's interest was something pursued by a variety of service/middle-class groupings who sought to prevent extensive *new* office building, as well as providing more local private housing. Within the following decade, this strict conservation policy was extended, so that, by 1985, 2000 buildings had been listed as having special architectural or historical merit (more than in the six other localities put together). The local authority also gave extensive grants to the owners of the town's private Regency housing stock for improvement and renovation. None of this was by any means inevitable. It was because of strong pressure from middle-class groupings that the council was encouraged to develop a local strategy which, as it turned out, has been extremely successful in encouraging new firms to relocate (especially their headquarters) in various Regency mansions, in attracting increasing numbers of visitors and in developing up-market shopping centres. It is worth noting how much smaller has been the level of spending on local employment initiatives compared with that on Regency restoration (less than one-twelfth), although some of the funding for restoration has derived from central government sources. Cheltenham has benefited enormously from the fairly recent reassertion of the value of 'heritage' and especially from the fact that Regency is probably the most highly valued of all British architectural styles. Cheltenham provides a good setting for the development of an 'educationally oriented holiday market', perhaps the best context possible given that it does not possess a university. We might almost see Swindon and Cheltenham as complementary symbols of Conservative Britain. Swindon is the place where 'hitech' industry is allowed to rip and where workers live (and increasingly vote Conservative) and Cheltenham is the place that is classically conserved for private sector managers/professionals to live in, where the state is represented (by GCHQ) and which workers may visit and be educated in (in modest numbers!). Neither town has a university or a polytechnic!

Cheltenham thus seems to have succeeded almost in spite of itself; and it succeeded because its social composition provided it with the social base of a strong and vociferous conservation movement. A partly parallel story can be told about Lancaster (City). Here, however, the development of a strong conservation movement was something that followed from the university being sited in Lancaster (against competition from a variety of places including Swindon and Cheltenham), as well as a college of education, and from the expansion of the hospital sector, all in the early 1960s. The town was rapidly 'embourgeoisified' (it had a heavily 'working-class' occupational structure in the 1950s) as manufacturing employment dramatically declined. The initial strategy pursued by a Conservative administration was at the time innovative, and consisted of encouraging the growth of small manufacturing firms by the provision of 'seedbed' premises. This strategy resulted from the fact that: large redundancies in a major company, following a merger, had created the public legitimacy of intervention, because it was believed that 'something needed to be done'; the perception that national regional policy could only work to the detriment of those areas such as Lancaster that did not at the time qualify for assistance; the legacy of a lack of local political conflict and supposedly limited councillor ability, which then allowed officers to formulate and implement new interventionist policies; the legitimation of such policies because of the development of professional groups locally who were committed to economic intervention; and the legacy of land and buildings in the City centre suitable for conversion.

The strategy was quite successful, especially in the later 1970s. However, new initiatives should have been developed during this period. The council was successful in attracting the support base for the Morecambe Bay gas field in the early 1980s, but by then the collapse of employment in manufacturing from 1981 onwards was undermining any hope that the plethora of new small manufacturing firms would be able to offset such decline. One possible initiative – city centre retailing expansion – only came on to the political agenda in the mid-1980s. Until then, it had been resisted, partly because of opposition by the strong conservation lobby, including a green Labour Party. Such a movement had already had some impact on the appearance of the city, thus facilitating the emergence of the second strategy pursued – the development of tourism. Rather like Cheltenham, the possibilities of this had not been readily appreciated until, in Lancaster's case, the later 1970s. Council officers began to conceive of Lancaster City as fitting in well with the changing external condition, i.e. the emergence of heritage tourism, in which the houses and workplaces of the industrial era became major sites for tourist development. It had previously been thought that the other half of the conurbation, Morecambe, was where tourism was concentrated. However, as tourist patterns were shifting away from 'modern' resorts to nostalgic symbols of the past, this permitted Lancaster City to be represented as a centre for heritage tourism (Urry, 1988). There was relatively little opposition to this from within Lancaster City, but plenty from Morecambe. It was thought inappropriate to spend money on facilities in Lancaster, rather than on, say, entertainment complexes in Morecambe. The forcing through of a tourism strategy for Lancaster has been in the face of considerable opposition. The efforts by the city council to upgrade facilities in Morecambe have not been very successful and a marked Lancaster–Morecambe split has developed. It is difficult to conceive a viable tourism strategy for the whole locality.

There are considerable parallels between the situations in Morecambe and Thanet. In each case there has been a Conservative council, which is expected by tourist-related interests both to keep rates low and to spend extensively on facilities that are specifically for potential visitors (not for local residents). This necessitates a kind of 'municipal Conservatism'. However, the tourist interest in such places is highly fragmented and localized, with very few major companies involved. There is thus little agreement on what might be thought desirable, and a limited ability to engage in long-term strategic planning for the locality.

In Thanet in the post-war period two important strategies have been pursued. First, there was the manufacturing strategy involving the building of industrial estates from the late 1950s. And, secondly, there has been the developing tourism/transport strategy, involving resort promotion, the provision of entertainment facilities, the attraction of new hotel investment, and the developments of the ferry service to Dunkirk, the Hoverport at Ramsgate and port expansion, also at Ramsgate.

Regarding the development of industrial estates, important external conditions in the 1960s were, economically, the availability of footloose industry, and politically, the development of regional policy. The main resource was the historical legacy of a labour force used to working for low wages in a non-union environment. This was not, however, a hegemonic project, partly because there was no tradition of manufacturing in Thanet and partly because there was a clear need to sustain a reasonably attractive environment for visitors. Also, the locality was rather poorly placed in terms of communications and educational resources.

For the tourism/transport strategy, there had always been fairly high expenditure on tourism-related activities. This is because the privately owned firms operating locally lacked the legitimacy or resources to provide collective or infrastructural facilities. However, there is more or less continuous resentment of the council's provision of such services, a situation rather similar to that found in Morecambe. There is little legitimacy for council involvement here, even in the case of publishing one single guide for the area. The council is taken as a scapegoat for the problems faced by the long-term decline of resort tourism. In order to atract a new hotel to Ramsgate, a major coup, the council had to take out shares in the private company rather than receive any money directly for the land that was involved. Given the general lack of interest in investment and the limited finance available, the council's organizational services are extremely limited. It has a weak bargaining position vis-à-vis private investors.

Partly in response to these different conflicts within the ruling Conservatives, alternative directions to local strategy have been advocated. One group has proposed considerable council intervention, particularly to develop the port at Ramsgate, while the other group has been anti-interventionist and pro-privatization. The extraordinary series of scandals, reflecting on the quality of the elected councillors, has almost removed all legitimacy from the council. This has been reinforced by the current electoral stalemate. The lack of legitimacy also partly explains the inability of the council or the local MPs to lobby effectively until recently, although the fact that Thanet, like Morecambe, did not fall into any of the traditional types of 'deserving locality' has made this difficult to achieve. It would seem that the development of the Channel Tunnel is a major resource for the whole of Kent and may ensure that, in future, even the more deprived parts in the east of the county are able to lobby successfully for a greater share of new schemes and resources. And there is, locally, an inherited tradition of intervention that may be of assistance. What there is not, however, is either any evidence of inspired strategic thinking among many local councillors or officers, or much possibility of local level bodies being so well-placed that the decline of Thanet can be seriously reversed.

3 Some concluding comments

On the basis of the experiences of these seven places, the following conclusions can be drawn:

(a) Up to the end of the 1970s a number of localities in both the North and South had been able to pursue 'public' strategies that had been reasonably successful in sustaining employment and income levels higher than would have otherwise occurred. Localities, therefore, did seem to make a particular difference and the financial cost was not always high.

(b) In the 1980s the combination of recession, the further internationalizing of the economy, and the restrictions on local government finance undermined most such strategies devoted to bolstering large-scale manufacturing employment, except in the particular case of Swindon. 'Private' strategies via 'central government localism' are now more common and the most that many localities can expect.

(c) In most cases where a 'successful' public strategy has been pursued for a period, this was normally initiated by strategically thinking officers (sometimes plus councillors) under pressure from well-informed and active public opinion which had been convinced that intervention was necessary – so forming a 'hegemonic strategy'.

(d) The resources available within different localities vary enormously and more attention should be paid to analysing how they can be utilized. This is something that can be done effectively by public planning authorities, although they need to do so in such a way that account is taken of the interests and beliefs of the many different social groups in the locality, in order to avoid the presumption of 'party chauvinism' found in all local political parties (Gyford, 1985).

(e) A major role in the development of the 'Regency-strategy' in Cheltenham and the 'tourism-strategy' in Lancaster was played by strongly organized middle-class groupings, who expressed their early opposition to the international style of modern architecture and have sought to conserve buildings constructed in local styles. To the extent to which they have been successful they provided those two localities with what has been an invaluable resource in their re-presentation as 'heritage' localities (see Dickens, 1988, more generally).

(f) The competition between councils to attract footloose firms is often wasteful, and there is a good case, therefore, for a county or regional authority to coordinate the local economic policies of lower level councils. It should be possible for councils (at least within a region) to avoid bidding each other up in terms of the incentives offered to footloose firms/plants.

(g) The effects of varied policies have been to produce quite considerable social differences between the seven localities (Cooke, 1989). This is not merely a question of differing rates of registered unemployment (although, of course, this has been spectacular in Liverpool and Middlesbrough), but much more generally of the availability of well-paid, interesting and skilled work for most of the population of a particular place. Cheltenham and, to a lesser extent, Swindon are the only two localities where this condition is more or less met. Places can also be compared in relation to the degree to which they fulfil people's cultural and aesthetic interests, interests that are highly stratified by class, gender, race and age. We could surmise that localities that are being 'done up' for tourists would be pleasanter for most residents but certainly not for all, and may certainly generate high levels of resentment.

Although the strategies pursued in these localities have hardly been 'progressive' in their consequences, they do show that localities *can* make a difference and that there can be an important role for elected public bodies. However, if in future such public strategies became viable again on an extensive scale, more attention needs to be devoted to the development of locally based policies that are simultaneously *professional* (that is, well-planned and well-organized, and involving efficient delivery of the service); *participatory* (that is, involving public debate and involvement to empower the local population, not merely to mobilize their support); and *productive* (in the sense that the activity in question meets some need, either as expressed through the market or through other socially acceptable means, which may have redistributive consequences) (see Gyford, 1985, on some of these dilemmas for the 'left'). It will be necessary to determine just how these contradictory objectives can be simultaneously met within particular localities.

But further 'local autonomy' cannot be the only principle operating here, because that would have the effect of magnifying inequalities between areas (for example, between Knowsley and Cheltenham). A further issue is how to locate localities within a regional or national system of resource *redistribution* between areas with very different resources, which does not entail the removal of powers of local decision-making, to conceive, plan and implement strategies for economic and social regeneration of each locality. This matter awaits resolution, although the recent experiences of 'local socialism' gives some important clues (Gyford, 1985; Mackintosh and Wainwright, 1987).

Indeed, there is a further even more basic issue, which concerns the degree to which localities are in any sense autonomous and what kind of locally specific coalition can be formed that is able to effect successful intervention. In the USA the combination of (a) the exceptional degree to which land is a free disposable commodity, (b) the considerable self-sufficiency of localities, and (c) the extensive authority of local officials makes each 'locality' distinctive and a major determinant of the life-chances of residents (Logan and Molotch, 1987). Cities, it is argued, have become 'growth machines' devoted to the increase of aggregate rents through the intensification of land use. Re-growth coalitions have been of great importance in American politics in the post-war period, helping to shape and direct locally the exercise of power. Particularly important in such coalitions in the USA and, indeed, elsewhere are those social groupings with an interest in the protection and enhancement of the immobile physical and social infrastructures. Harvey (1985) talks of how extensive cross-class alliances can develop within the locality, alliances that engage in locality or civic 'boosterism'. Central, he argues, to such a policy is the existence of a growth coalition able to use its political and economic power to push the locality into an upward spiral of perpetual and sustained accumulation. Such a coalition has to act as a kind of 'collective entrepreneur'; without it, localities may get 'left behind, stagnate, decay or drift into bankruptcy' (Harvey, 1985, p. 158).

The danger is that in late-twentieth-century Britain there are inhospitable 'conditions', combined with insufficient 'resources', which will mean that it will be well-nigh impossible to devise locally specific policies, at least partly under the sway of local authorities, to be able to prevent many localities getting seriously 'left behind'. This book has been an attempt to detail what the conditions and resources are that can permit the development of well-organized local interventions.

References: Chapter 9

Bagguley, P., Mark Lawson, J., Shapiro, D., Urry, J., Walby, S. and Warde, A. 1990. *Restructuring: place, class and gender*. London: Sage.

Cooke, P. (ed.). 1989. *Localities*. London: Unwin Hyman.

Dickens, P. 1988. *One nation? Social change and the politics of locality*. London: Pluto.

Fothergill, S. and Gudgin, G. 1982. *Unequal growth*. London: Heinemann.

Gyford, J. 1985. *The politics of local Socialism*. London: Allen and Unwin.

Harvey, D. 1985. *The urbanization of capital*. Oxford: Blackwell.

Laffin, M. 1986. *Professionalism and policy: the role of the professions in central–local government relationship*. Aldershot: Gower.

Larson, M. S. 1977. *The rise of professionalism*. Berkeley, Calif.: University of California Press.

Logan, J. R. and Molotch, H. L. 1987. *Urban fortunes. The political economy of place*. Berkeley, Calif.: University of California Press.

Mackintosh, M. and Wainwright, H. (eds). 1987. *A taste of power. The politics of local economics*. London: Verso.

Murgatroyd, L., Savage, M., Shapiro, D., Urry, J., Walby, S. and Warde, A. 1985. *Localities, class and gender*. London: Pion.

Rhodes, R. A. W. 1988. *Beyond Westminster and Whitehall*. London: Unwin Hyman.

Urry, J. 1988. 'Cultural change and contemporary holiday-making,' *Theory, Culture and Society*, 5, 35–55.

Warde, A. 1985. 'The homogenisation of space? Trends in the spatial division of labour in twentieth-century Britain.' In H. Newby, J. Bujra, P. Littlewood, G. Rees and T. Rees (eds). *Restructuring capital. Recession and reorganisation in industrial society*. London: Macmillan.

Index

N.B. * indicates heading also covered in several locality chapters